MEAP Mathematics Coach

Grade 4

Jerome D. Kaplan, Ed.D.
Professor Emeritus
Seton Hall University

A division of Haights Cross Communications.

For information about permission, write to:
Educational Design, 345 Hudson Street, New York, NY 10014-4502

ISBN 1-58620-082-8 EDI 303

10 9 8 7 6 5 4 3 2

TABLE OF CONTENTS

Mathematics Content Standards and Benchmarks

TO THE STUDENT

This book is called ***MEAP Mathematics Coach, Grade 4***. It will help you prepare for the MEAP Test for Grade 4.

Here is how ***MEAP Mathematics Coach*** can help you:

- It shows you what the math questions on the MEAP Test are like.
- It tells you what you need to know to do well on the test.
- Finally, it gives you practice using the math skills you will need for the test.

The MEAP Test in Math has many **multiple-choice** questions. They are like the ones you will work with in this book. After each question there are four possible answers. Only one is correct. The others are wrong. You must mark the one **correct** answer for each question.

The MEAP Test also has **open-ended** questions. On these questions, you will have to write a short sentence, or draw a diagram, or give a full explanation of why something works. You will find open-ended questions after the Strand Reviews.

Here are some tips that will help when you work in this book and take the test:

✓ **Read each question carefully.**

✓ **Work as carefully as you can.**

✓ **Make sure you answer the question that is asked.**

✓ **Ask yourself if the answer makes sense.**

✓ **On multiple-choice questions, if you cannot decide on the answer, make the best guess you can. There is no penalty for guessing.**

✓ **Answer as many questions as you can.**

✓ **On open-ended questions, answer the questions as completely as you can.**

Use these tips throughout the book and when you take the test.

LESSON 1

IDENTIFYING PATTERNS

EXAMPLE 1

The shaded parts in these four rectangles represent fractions. As you go from rectangle to rectangle, how much does each fraction increase?

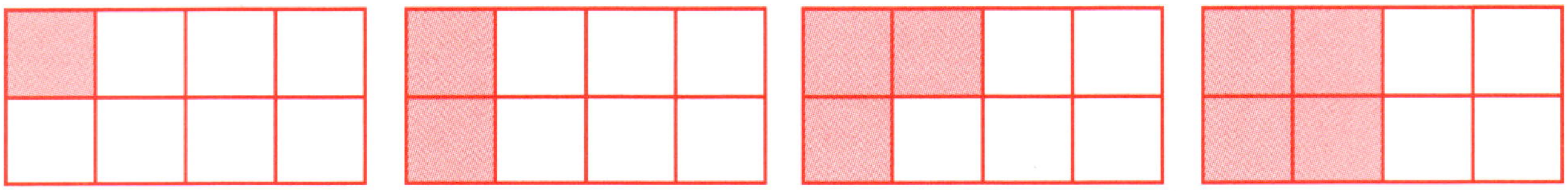

1 2 3 4

STRATEGY: **Find the fraction for the shaded parts of each rectangle. Then compare the increase from Rectangle 1 to Rectangle 2, from Rectangle 2 to Rectangle 3, and from Rectangle 3 to Rectangle 4.**

STEP 1: List the fractions for the shaded parts.

Rectangle 1: $\frac{1}{8}$

Rectangle 2: $\frac{2}{8}$

Rectangle 3: $\frac{3}{8}$

Rectangle 4: $\frac{4}{8}$

STEP 2: Find the difference between the amount shaded in each rectangle.

SOLUTION: **The parts shaded increase by $\frac{1}{8}$ of a rectangle each time.**

EXAMPLE 2

What figure would replace the question mark in this pattern?

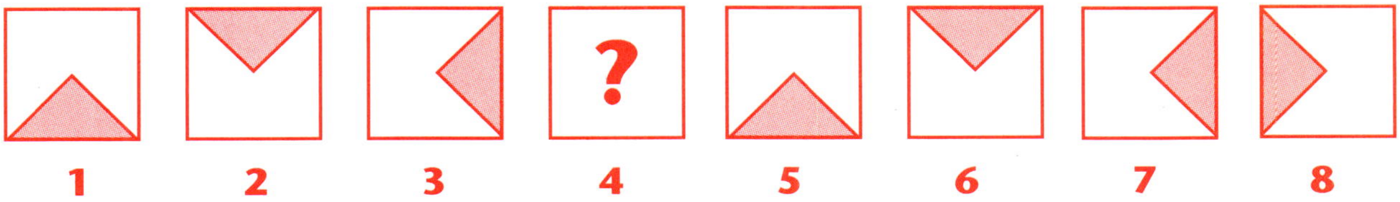

STRATEGY: **Find the rule that describes the pattern. Then apply the rule to Figure 4.**

STEP 1: Ask: How would you decribe Figures 1 and 2?

In Figures 1 and 2, the shaded triangles are opposites of each other.

STEP 2: How would you describe Figures 5 and 6? And Figures 7 and 8?

The shaded triangles are opposites of each other.

STEP 3: Describe the rule.

In each pair, the shaded triangles are opposite each other.

STEP 4: Apply the rule to Figure 4.

Look for a shaded triangle that is the opposite of the triangle in Figure 3.

SOLUTION: **B goes in Figure 4, because its shaded triangle is the opposite of the shaded triangle in Figure 3.**

SAMPLE TEST QUESTIONS

1 What would replace the question mark in this pattern?

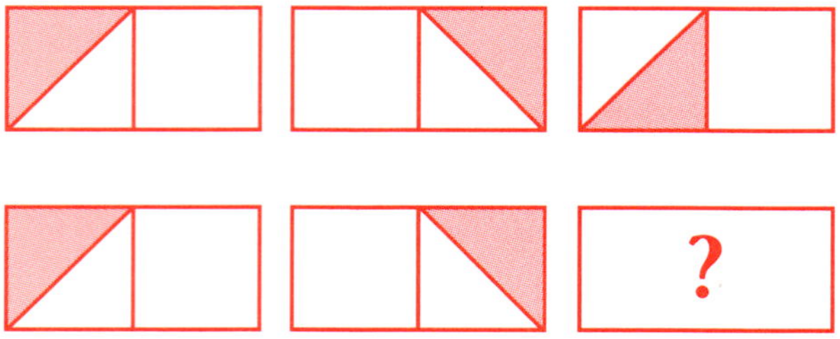

A 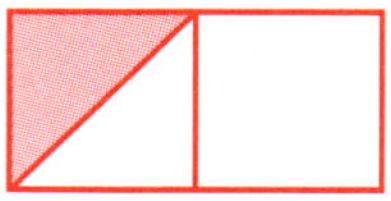

B 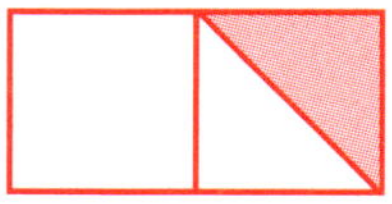

C 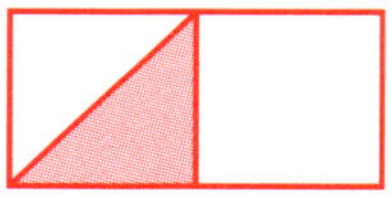

D 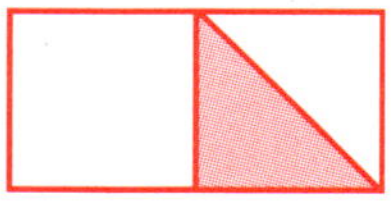

2 What symbol should replace the question mark in the pattern below?

✿✿▲▲▲◆✱✱✿ _?_ ▲▲▲◆✱✱

A ▲

B ✿

C ◆

D ✱

3 How many cubes will be needed to build Step 4?

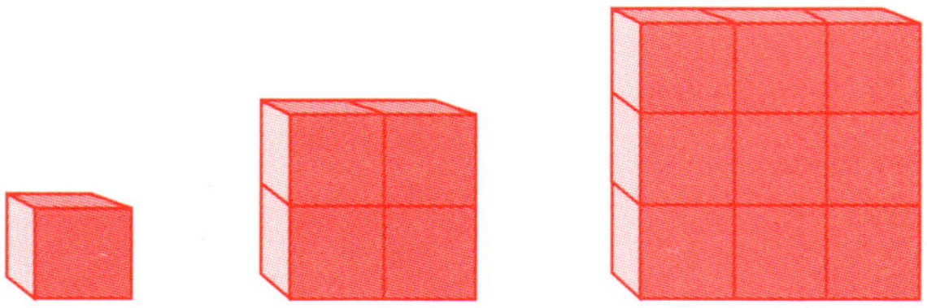

A 12

B 15

C 16

D 25

4 What is the next figure in this pattern?

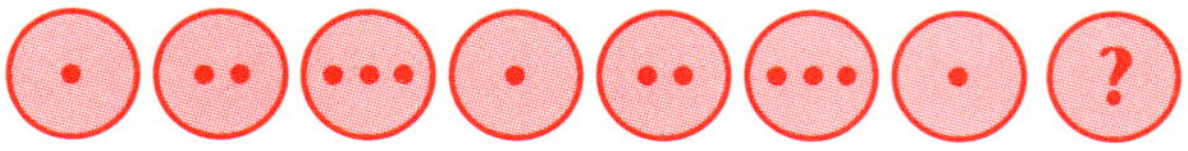

A 

B 

C 

D 

LESSON 2

IDENTIFYING NUMBER PATTERNS

EXAMPLE 1

What is the next number in this pattern?

20, 17, 14, 11, ___ , 5, 2

STRATEGY: **Follow these steps.**

STEP 1: Ask: Do the numbers get larger or smaller?

In this list, the numbers get smaller.

STEP 2: Ask: What number do you subtract from each number to get the next number in the pattern?

20 – 17 = 3

17 – 14 = 3

14 – 11 = 3

You subtract 3 from each number to get the next number.

STEP 3: Subtract 3 from 11 to find the missing number.

11 – 3 = 8

SOLUTION: **The missing number is 8.**

EXAMPLE 2

What is the next number in the pattern?

1, 2, 3, 5, 8, 13, 21, ___

STRATEGY: **Look for an unusual pattern.**

(In this pattern, the numbers get larger. But you can't add the same number each time to get the next number! You'll have to try something different.)

STEP 1: Try adding the first two numbers: 1 + 2 = 3.

You get the third number, 3.

STEP 2: Try adding the second and third numbers: 2 + 3 = 5.

You get the fourth number, 5.

STEP 3: And so on.

The pattern is: add two numbers to get the next number.

STEP 4: To get the eighth number, add the sixth and seventh numbers.

13 + 21 = 34

SOLUTION: **The missing number is 34.**

SAMPLE TEST QUESTIONS

1 What is the missing number in the number pattern?

6, 14, 22, 30, 38, ___ , 54, 62

A 42

B 46

C 48

D 50

2 Which two numbers would be next in the pattern?

6, 11, 16, 21, ___ , ___

A 25, 30

B 26, 31

C 27, 32

D 28, 33

3 What is the missing number in this pattern?

160, 80, 40, 20, ___ , 5

A 15

B 12

C 10

D 8

4 What number comes next in this pattern?

0, 1, 1, 2, 3, 5, 8, ___

A 11

B 12

C 13

D 14

5 What number comes next in this pattern?

10, 11, 10, 12, 10, 13, 10, ___

A 14 **B** 13

C 12 **D** 10

6 Tammy's father saved $100 in January. Every month after January he saved $200 more than he had in each previous month. How much money had he saved by June 1?

A $3,600

B $2,500

C $1,500

D $900

LESSON 3

EXTENDING GEOMETRIC PATTERNS

EXAMPLE 1

What shape comes next in this pattern?

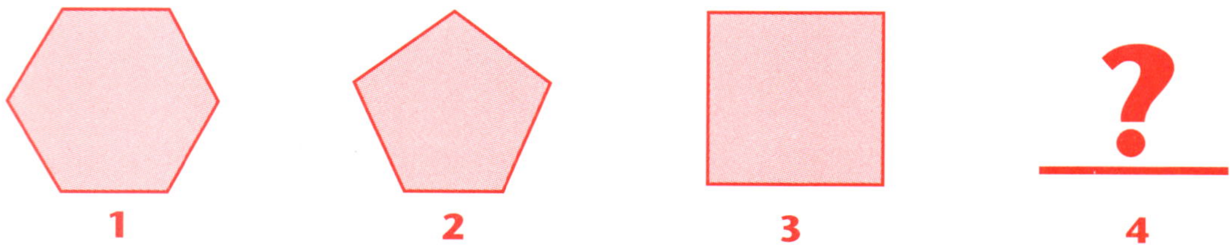

STRATEGY: **To find the pattern, count the sides.**

STEP 1: Ask: How many sides does each figure have?

Figure 1 has 6 sides.

Figure 2 has 5 sides.

Figure 3 has 4 sides.

STEP 2: Write the number of sides in a list.

List: 6, 5, 4

STEP 3: Find the next number in this pattern.

The next number is 3.

STEP 4: Ask: What figure has 3 sides?

SOLUTION: **A triangle has 3 sides—it is the next figure in the pattern.**

EXAMPLE 2

The figure below is a square. Which group of segments would be used to make a square?

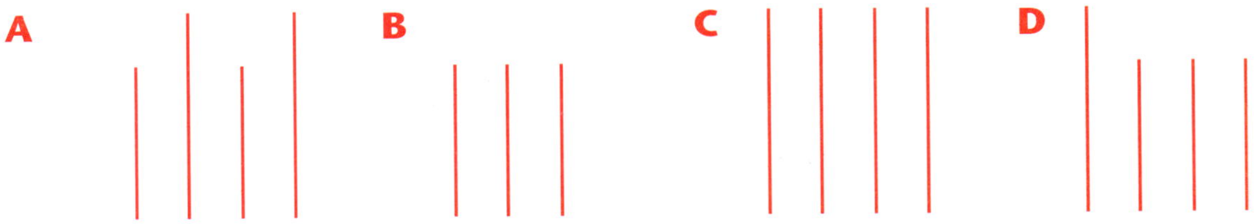

STRATEGY: **Pretend that the square is made up of sticks that you can move around, so that you don't have to worry about which way they are facing. Then see which group of segments they match.**

STEP 1: Ask: How many segments are used to make a square?

Four.

STEP 2: Ask: What do you know about the lengths of these 4 segments?

They are all the same length.

STEP 3: Look for a group of segments that matches your description of the segments that make up a square.

SOLUTION: **C is the correct answer—it's the only one that has 4 segments, all the same length.**

EXAMPLE 3

If the pattern continues, the black dot in the next graph will be at which coordinates?

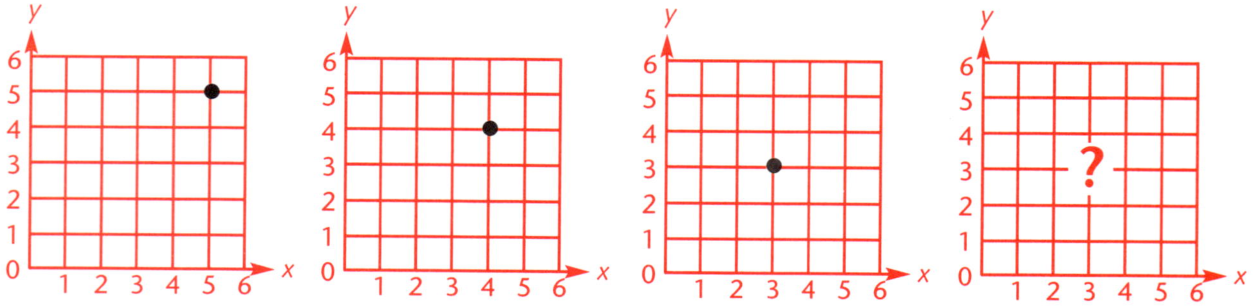

STRATEGY: **Look for a pattern in the dots.**

STEP 1: Write down the ordered pairs for the three dots. Remember that you always write the number for the horizontal (left-to-right) axis first.

The three ordered pairs for the three dots are: (5,5), (4,4), (3,3).

STEP 2: Figure out the pattern.

With each graph, both numbers in the ordered pair get smaller by one.

STEP 3: Write the next ordered pair in this pattern.

SOLUTION: **The black dot in the next graph will be at (2,2).**

SAMPLE TEST QUESTIONS

1 Which of the following comes next in the pattern?

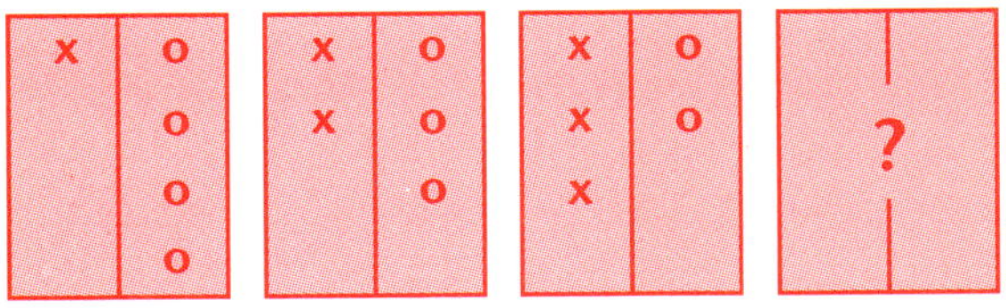

A
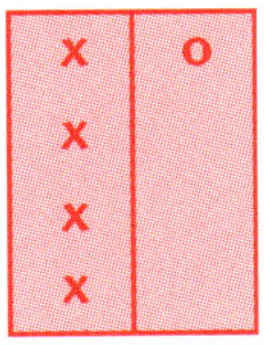

B
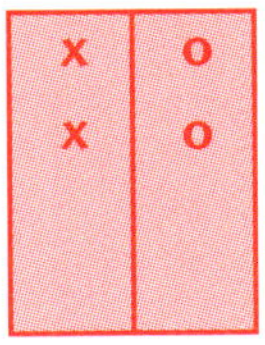

C
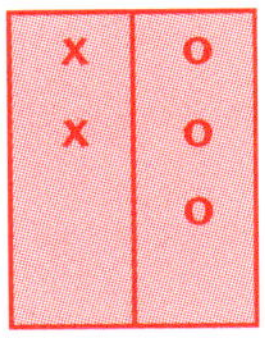

D
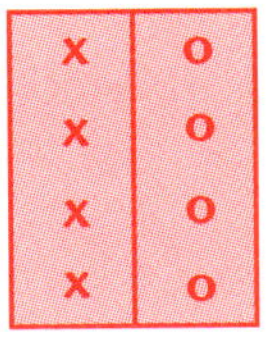

2 How does shape X compare to shape Y?

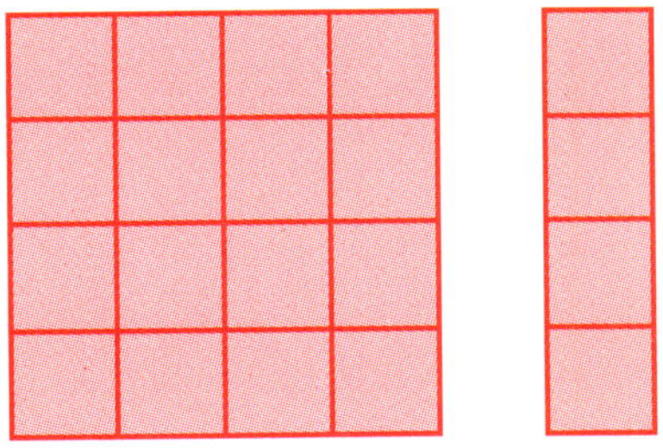

X Y

A They are the same size.

B X is twice as big as Y.

C X is 3 times as big as Y.

D X is 4 times as big as Y.

3 Which is the next picture in this pattern?

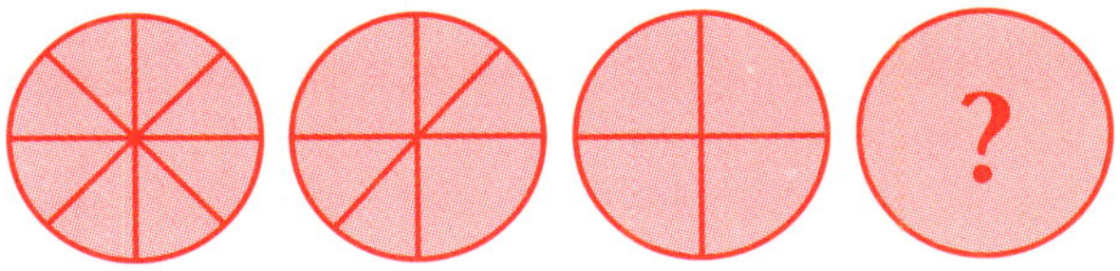

A
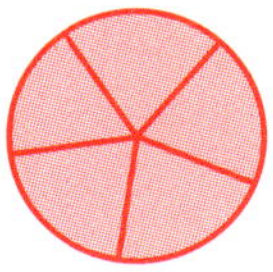

B
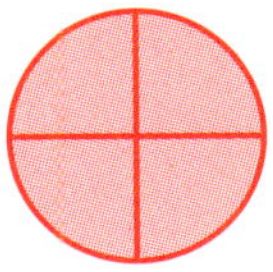

C
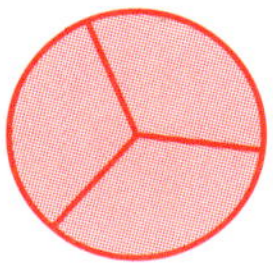

D
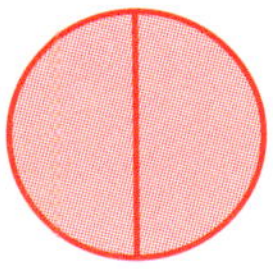

4 Which segments are used to make a figure like this?

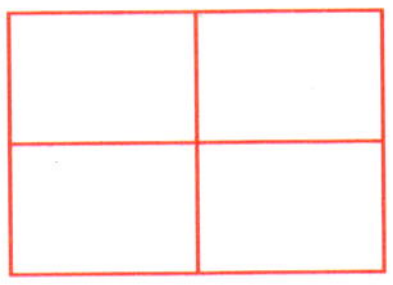

A

B

C

D

LESSON 4

EXTENDING PATTERNS OF ORDERED PAIRS

EXAMPLE 1

Which pair of numbers should come next in the column?

X	Y
1	6
2	12
3	18
?	?

STRATEGY: **Find the pattern in the X and Y columns.**

STEP 1: What is the next number in the X column?

The next number in the pattern 1, 2, 3, ... is 4.

STEP 2: What is the next number in the Y column?

The next number in the pattern 6, 12, 18, ... is 24. Count by 6's.

SOLUTION: **The pair that comes next is (4,24).**

EXAMPLE 2

Which pair of numbers would come next in the pattern?

(4,10), (8,18), (12,26)

A (13,27) **B** (14,32)

C (16,34) **D** (16,36)

STRATEGY: **Look at the numbers. The first number in each pair increases by 4. Add 4 to 12: 4 + 12 = 16. The second number increases by 8. Add 8 to 26: 26 + 8 = 34.**

SOLUTION: **C is the correct answer, (16,34).**

SAMPLE TEST QUESTIONS

1 What number fits in the blank space?

2	4	6	8
4	8	12	16
8	16	24	32
16		48	64

A 24

B 32

C 42

D 48

2 Which pair of numbers should come next in the column?

X	Y
17	21
20	24
23	27
?	?

A (26,30)

B (26,29)

C (25,30)

D (25,39)

3 If this pattern continues, what will be the next pair of numbers?

(4,15), (6,23), (8,31), (?,?)

A (10,4)

B (10,39)

C (10,38)

D (8,39)

4 The following is part of a hundreds board.

1	2	3	4	5	6	7	8	9	10
11	12	13	14	15	16	17	18	19	20
21	22	23	24	25	26	27	28	29	30

Below is a portion of the same board. What number should be in the box with the question mark inside?

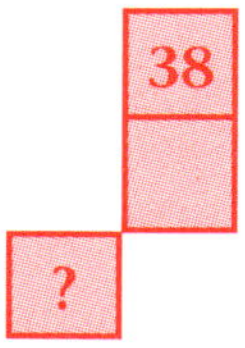

A 47

B 57

C 58

D 67

REVIEW OF LESSONS 1-4

1 Which figure comes next in this pattern?

A 

B 

C 

D 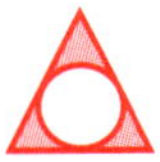

2 What number is missing from this pattern?

30, 26, 22, 18, ___, 10, 6

A 16

B 15

C 14

D 12

3 Which pair of numbers should come next in the column?

A	B
6	2
9	3
12	4
☐	☐

A (15,5)

B (15,4)

C (14,5)

D (14,6)

4 What is the next number of this pattern?

10, 21, 32, 43, ___

A 49

B 51

C 52

D 54

OPEN-ENDED QUESTIONS

1 Make your own geometric pattern using these figures. Explain your pattern.

2 a. Find the pattern for the pairs of numbers in the table.

X	Y
2	7
4	9
6	11
8	13

What is the pattern?

b. Use the same pattern to create your own table of number pairs. You may start with any number you like as long as you use the same pattern.

p	q

LESSON 5

TWO-DIMENSIONAL FIGURES

Here are several geometric figures you should know:

Line segment

A part of a line. It has a beginning and an end, and it is straight.

Angle

An angle is formed by two lines or two line segments that meet or cross each other. The point where the lines or line segments meet is the **vertex**. Every angle has a vertex.

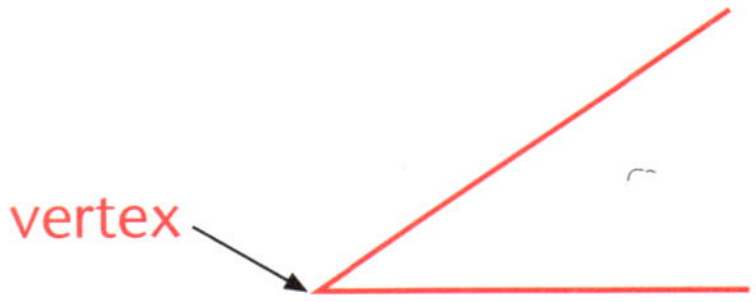

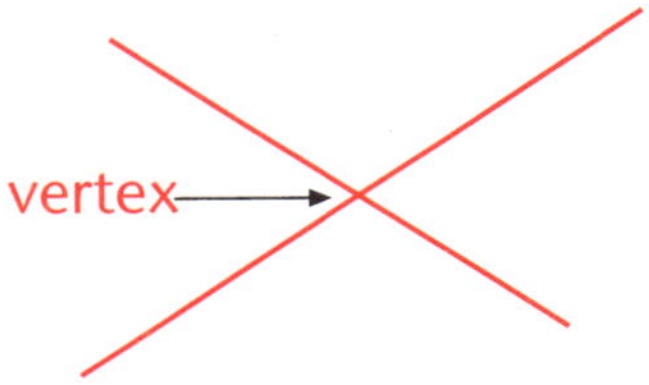

Triangle

A figure with three sides and three angles. All these figures are triangles.

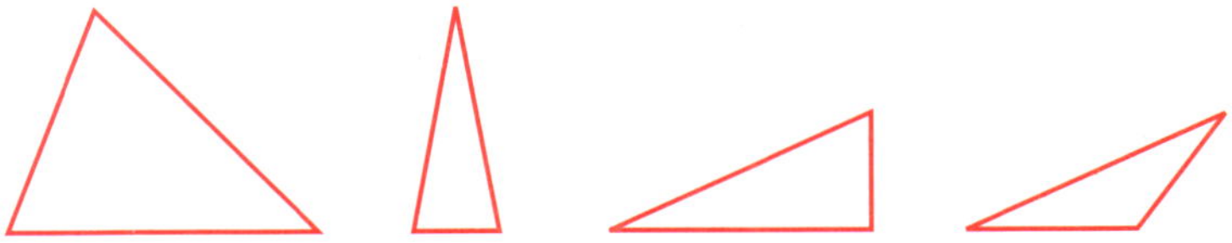

Rectangle

A figure with four sides and four square corners (four **right angles**).

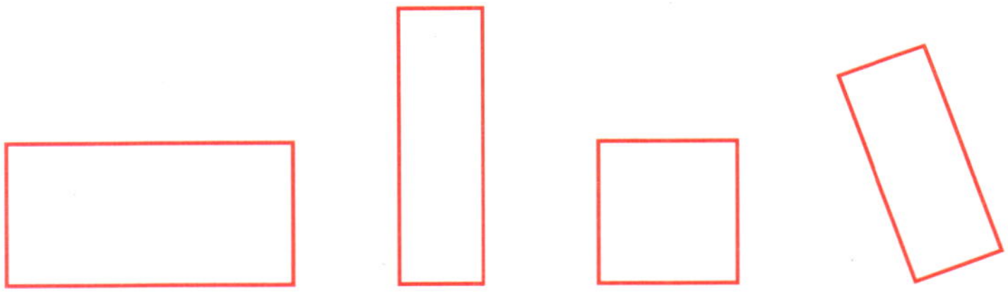

Square

A special kind of rectangle, with all sides equal.

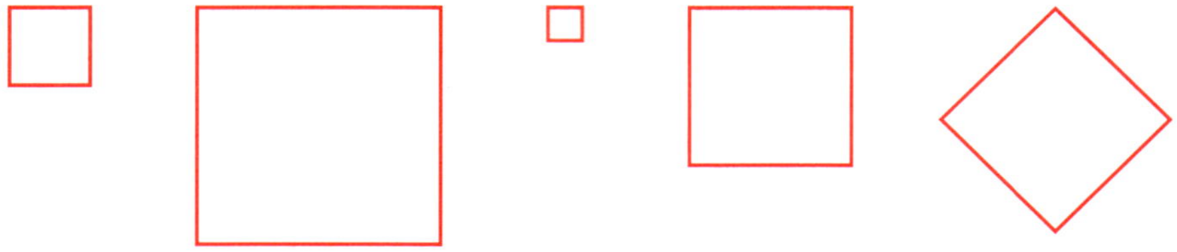

Circle

A round figure, like the one below. All points on a circle are the same distance from the center of the circle.

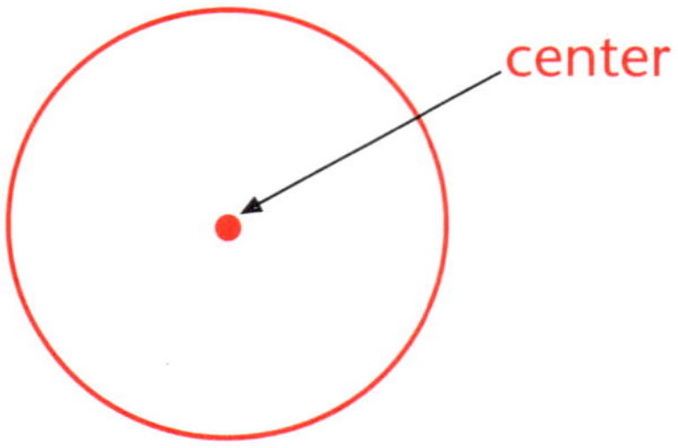

Polygon

A figure made up of straight sides. The figure is closed—there are no gaps or openings in it.

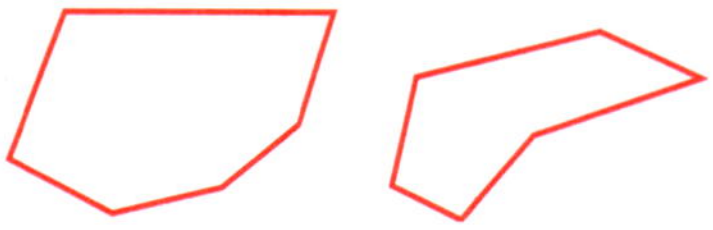

All triangles and rectangles are polygons.

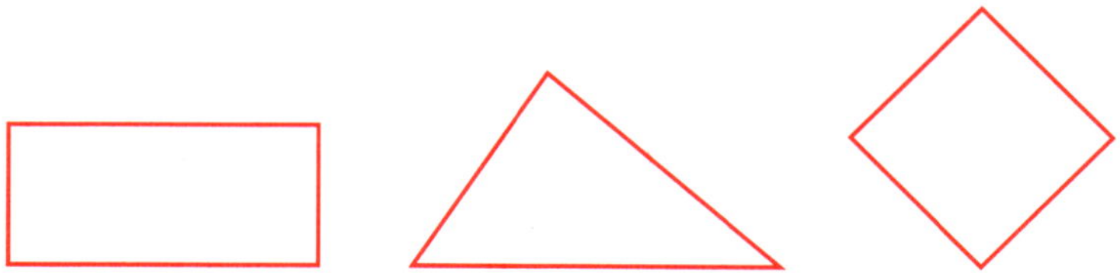

There are two other special types of polygons that you should know.

Pentagon A figure with 5 sides and 5 angles.

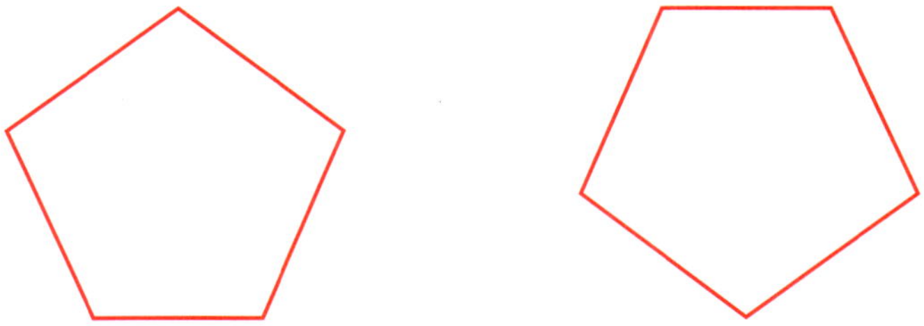

Hexagon A figure with 6 sides and 6 angles.

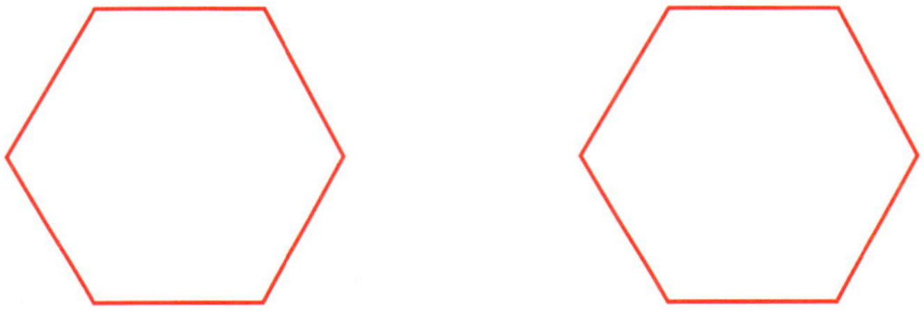

But not every figure is a polygon. Circles are not polygons. Neither are these:

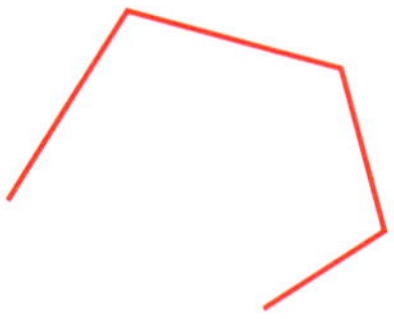

Not a polygon because it is not closed.

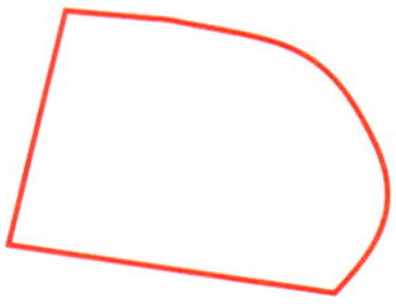

Not a polygon because it has a round side.

LINES OF SYMMETRY

If a figure can be folded so that its two halves match, then it has a **line of symmetry**.

This line is called the **line of symmetry.**

A line of symmetry divides a figure in half.

Figures can have more than one line of symmetry.

EXAMPLE 1

How many lines of symmetry does this rectangle have?

STRATEGY: **Try to fold the rectangle in half as many ways as you can.**

You can fold it in half two ways so that the halves match each other.

SOLUTION: **The rectangle has two lines of symmetry.**

CONGRUENT FIGURES

Two figures are **congruent** if they have the same shape and size.

These triangles are congruent.

Figures do not have to face the same way to be congruent. The two triangles below are also congruent.

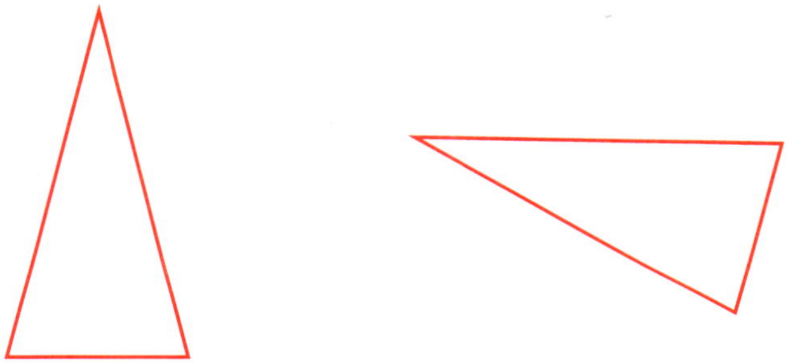

EXAMPLE 2

Which pieces of the puzzle are congruent?

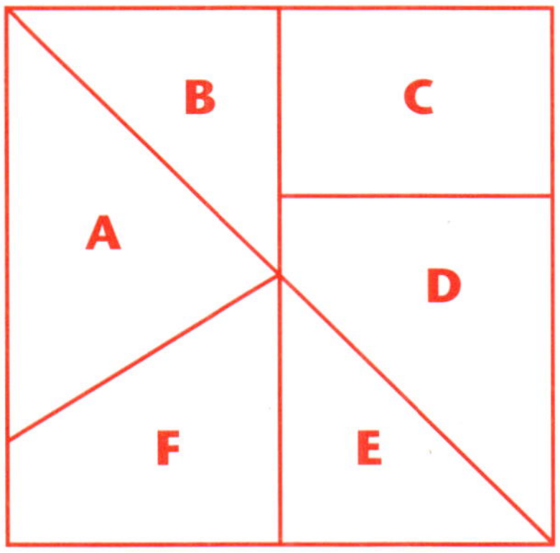

STRATEGY: **Remember what *congruent* means.**

Find two pieces that are the same shape and size.

SOLUTION: **Triangles B and E are the same shape and size.**

SAMPLE TEST QUESTIONS

1 Which of the following is NOT a polygon?

A 

B 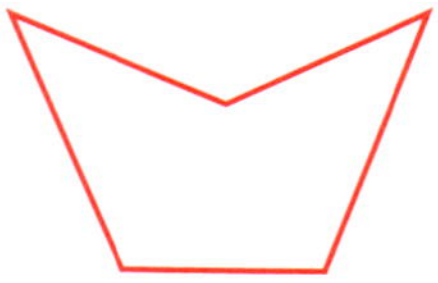

C 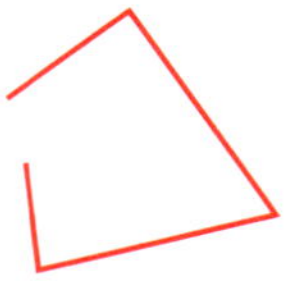

D 

2 Which figure has the most vertices (plural of vertex)?

A hexagon

B square

C pentagon

D triangle

3 Which number is inside the circle and outside the triangle?

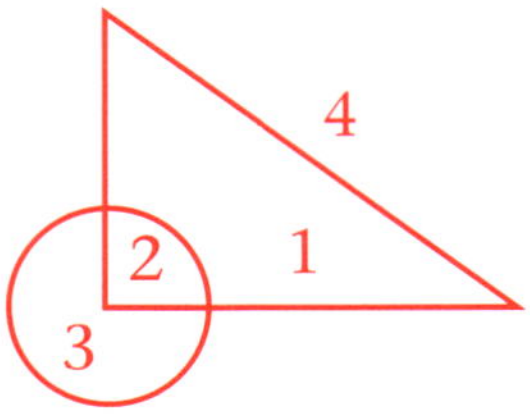

A 1

B 2

C 3

D 4

4 Which figure shows a line of symmetry?

A 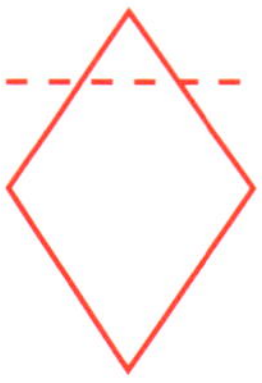

B 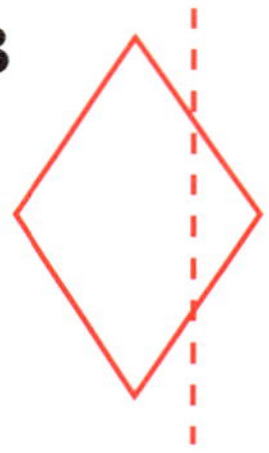

C 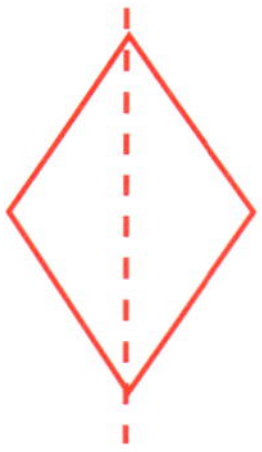

D 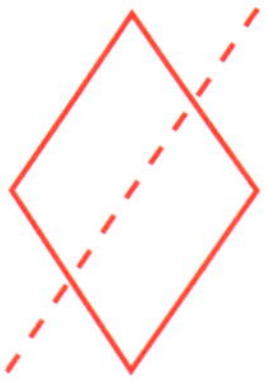

5 Look at the following figures. Which two are congruent?

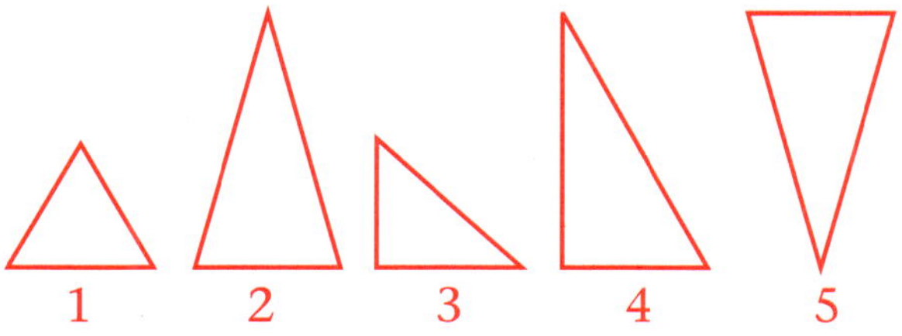

A 2 and 4

B 1 and 3

C 2 and 5

D 1 and 2

6 Which statement is NOT true?

A All squares are rectangles.

B Some rectangles are squares.

C All triangles are polygons.

D Some hexagons are not polygons.

7 What do the following figures have in common?

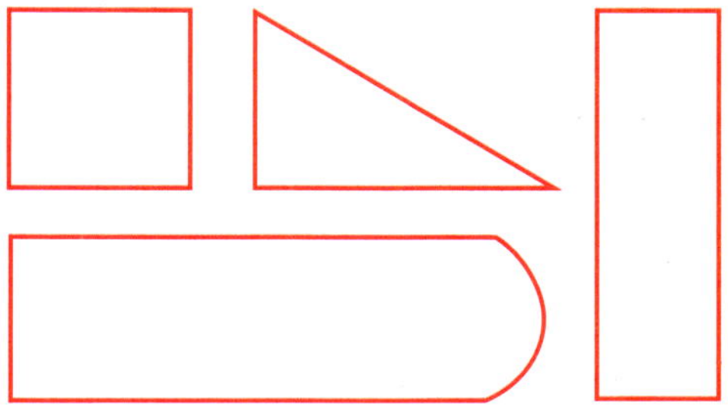

A They are all polygons.

B They all have at least one right angle.

C They all have 1 line of symmetry

D They have nothing in common.

8 Which statement is NOT true?

A Circles have many lines of symmetry.

B Two circles cannot be congruent.

C Two hexagons can be congruent.

D Some polygons are pentagons.

LESSON 6

THREE-DIMENSIONAL FIGURES

Three-dimensional figures are not flat figures. They have capacity (or volume). They are also called **solids**.

Rectangular Prism

Example: a box.

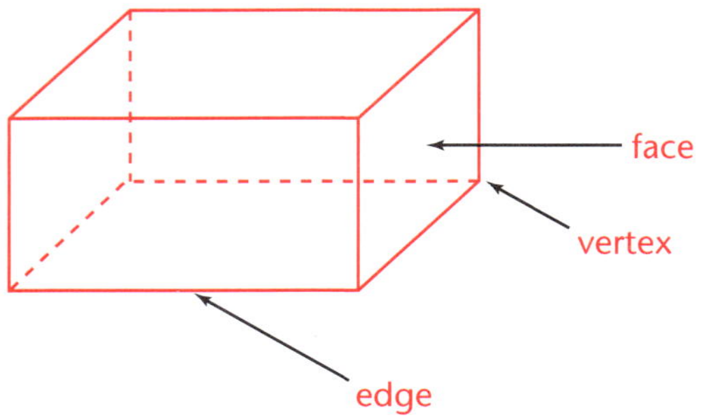

The flat surfaces are called **faces**.

The faces meet at **edges**. The edges are **line segments**.

The edges meet at **vertices** (plural of **vertex**).

Cube

Example: blocks that young children play with. A special kind of rectangular prism. All its faces are squares.

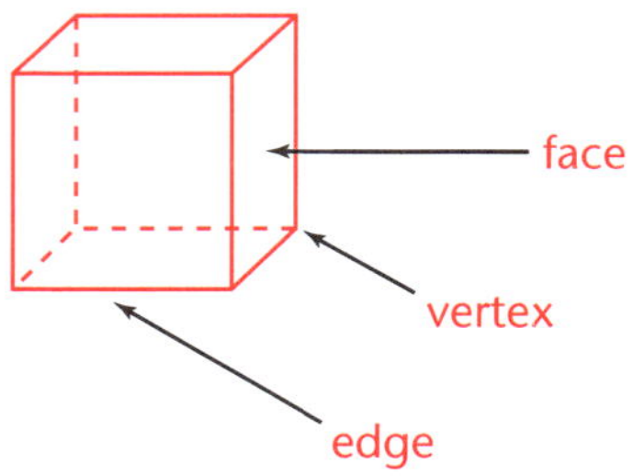

Pyramid

The base is often a square. The other faces are triangles.

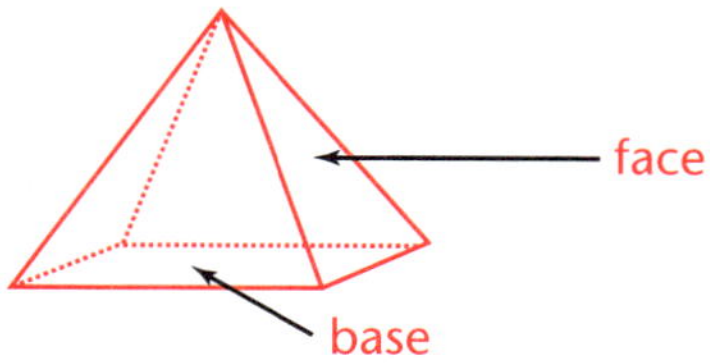

Cylinder

Example: a can. The top and the bottom faces are circles.

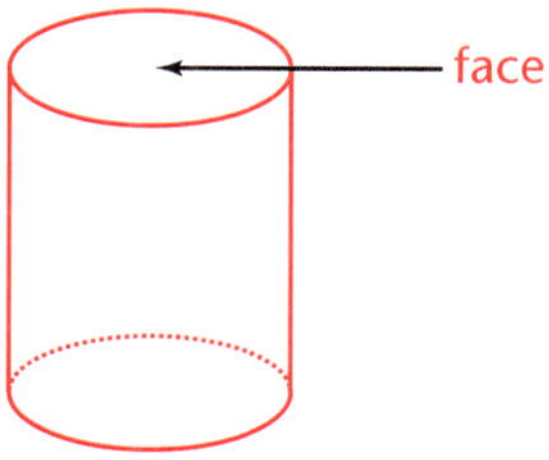

Cone

Example: an ice cream cone. It has one face, a circle.

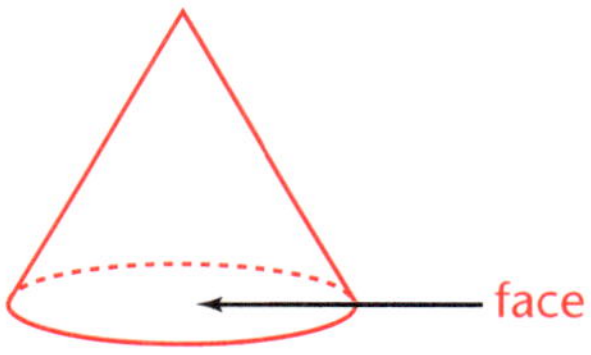

Sphere

Example: a ball. It has no faces.

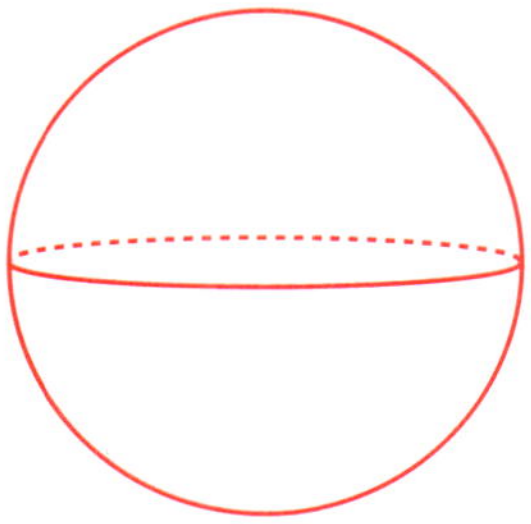

EXAMPLE 1

How many total edges does a square pyramid have? Look at the diagram on page 29.

STRATEGY: **Count the segments where two faces meet.**

The sides of the square:	4 edges
The segments that come together at top:	4 edges
Total:	8 edges

SOLUTION: **The square pyramid has 8 edges.**

EXAMPLE 2

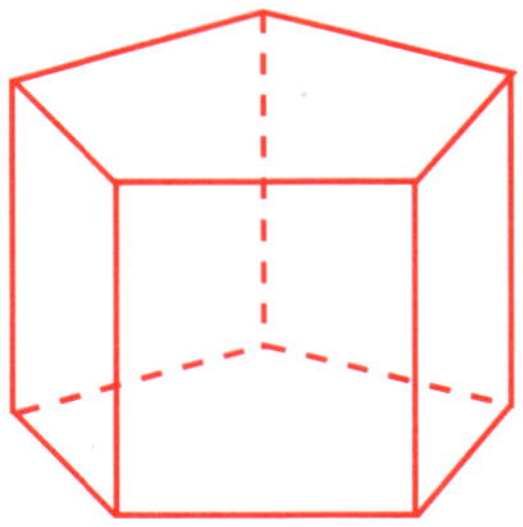

How many total faces does this figure have?

STRATEGY: **Count the faces.**

The top and the bottom:	2 faces
Rectangles on the sides:	5 faces
Total:	7 faces

SOLUTION: **The figure has 7 faces.**

SAMPLE TEST QUESTIONS

1 What is this figure called?

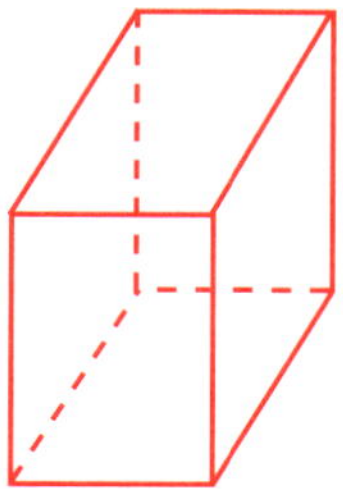

A rectangular prism

B cone

C pyramid

D cube

2 What shape are the faces of a cube?

A triangle

B square

C pentagon

D hexagon

3 How many total faces does this figure have?

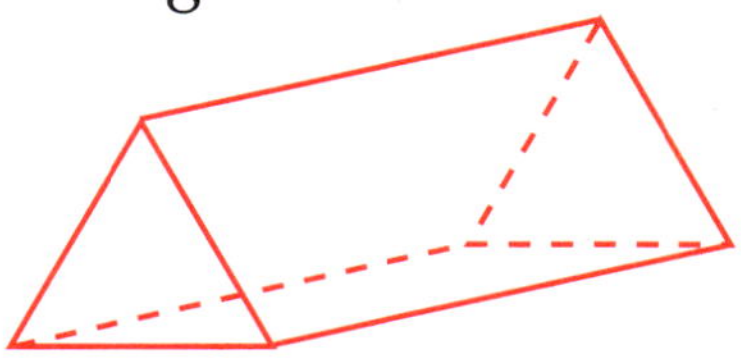

A 3 **B** 4

C 5 **D** 6

4 Which of these figures has no faces?

A cylinder

B pyramid

C cone

D sphere

5 How many more vertices than faces does this figure have?

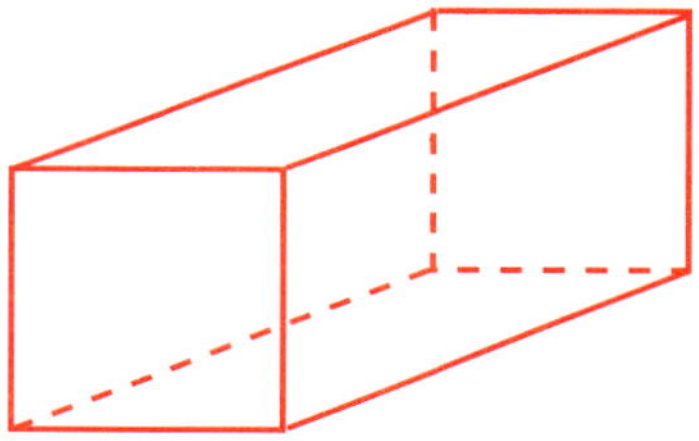

A 2

B 3

C 4

D 5

6 How many edges does a cube have?

A 8

B 12

C 14

D 16

LESSON 7

SLIDES, TURNS, AND FLIPS

There are three basic ways to move figures.

SLIDE

When you slide a figure across the page.

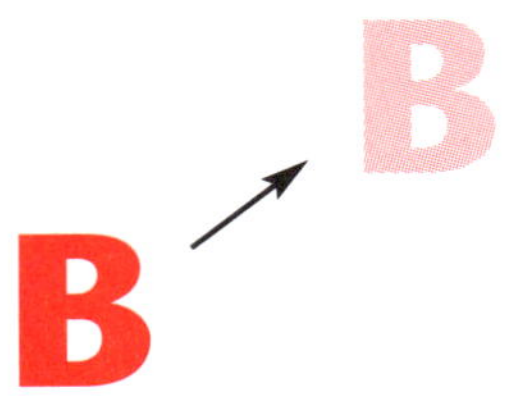

TURN

When you turn a figure around a point. A full turn completes a circle. The figure returns to the starting place. The turn shown here is a quarter turn, or one-fourth turn.

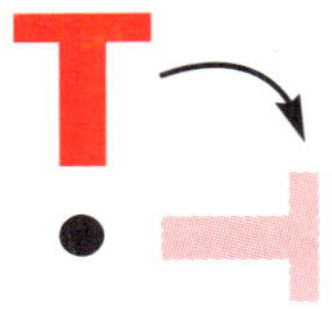

FLIP

When you flip a figure over a line.

EXAMPLE 1

If it is 5:15 on this clock, what time would it be if the hands on the clock were flipped to the left?

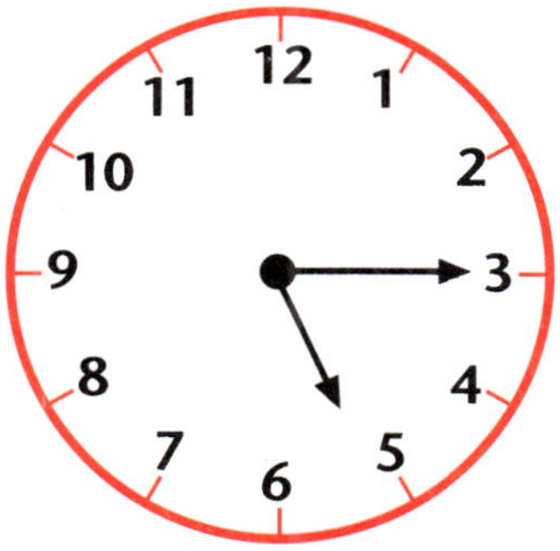

STRATEGY: **Make a drawing of the flip.**

STEP 1: Draw the line you are flipping the hands over.

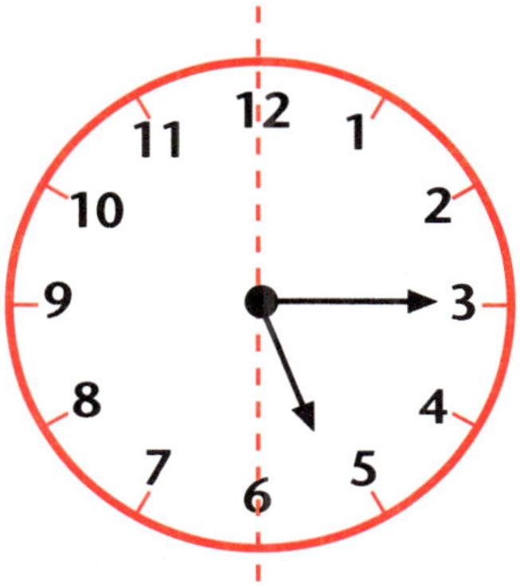

STEP 2: Draw the flip of the hands of the clock. (Remember: Each hand will now be just as far to the left as it had been to the right.)

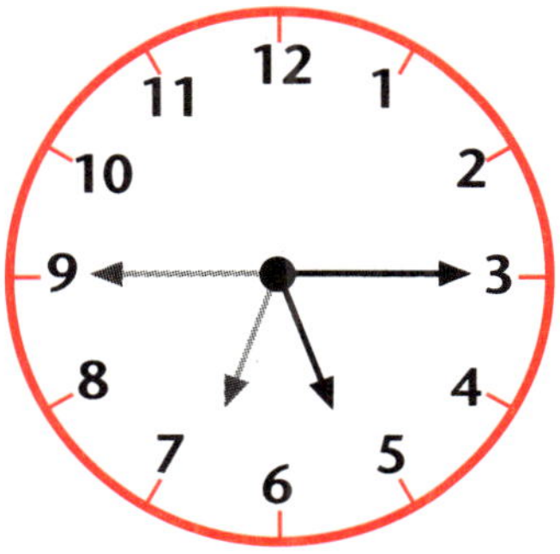

STEP 3: Read the time from the flipped hands. (The flipped hands are in grey.)

SOLUTION: **The time is 6:45.**

EXAMPLE 2

Suppose you want to slide this figure 2 units down. What would point A then be?

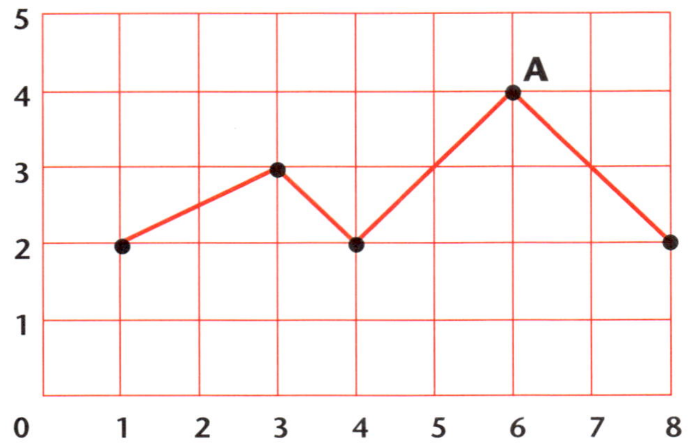

STRATEGY: **Use your finger to find the point.**

STEP 1: Find A on the figure. It is at (6,4).

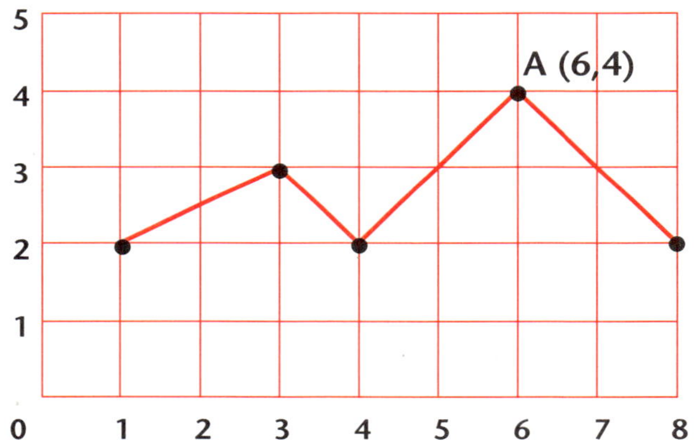

STEP 2: Slide A down 2 units.

Put your finger at the place 2 units down from (6,4).

STEP 3: Name the point after you slide A 2 units down.

SOLUTION: **Point A shifted 2 units down would slide to (6,2).**

SAMPLE TEST QUESTIONS

1 Which of these shows a flip of the letter Y?

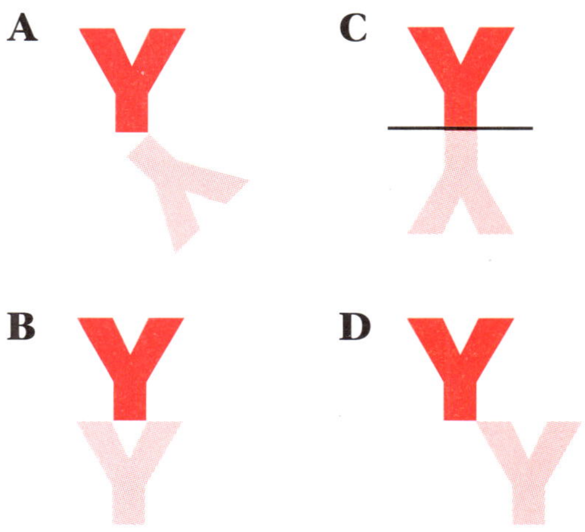

2 What kind of a turn is this?

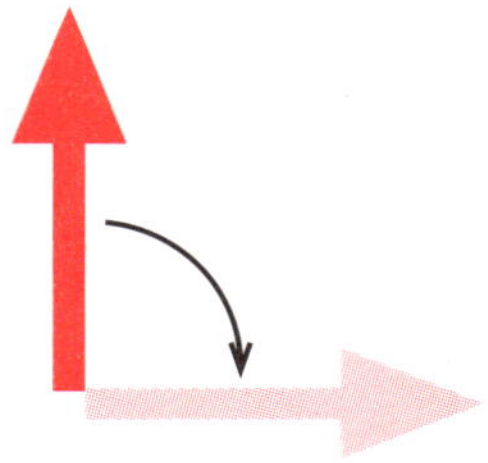

A full turn

B half turn

C quarter turn

D complete turn

3 To get the figure at X to move to location Y, what should you do?

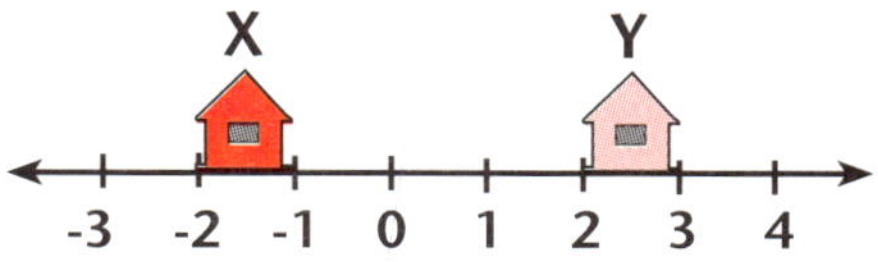

A Move figure X 4 units to the left.

B Move figure X 2 units to the right.

C Move figure X 3 units to the right.

D Move figure X 4 units to the right.

4 What do you have to do to move point Y to point V?

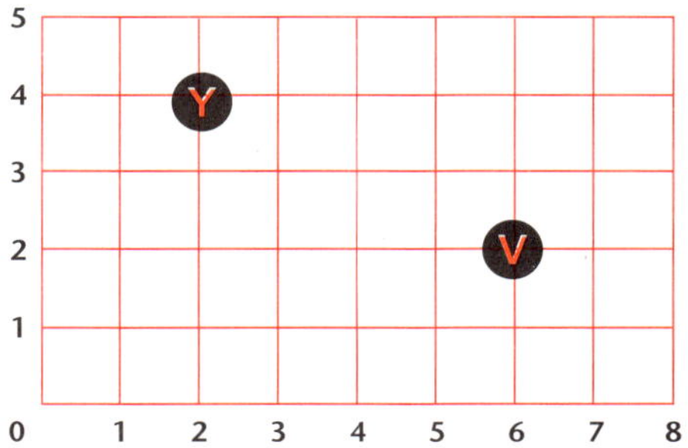

A Move Y 4 units to the right.

B Move Y 2 units to the right.

C Move Y 4 units to the right and 2 units down.

D Move figure Y 4 units down and 2 units to the right.

LESSON 8

PROPERTIES OF SEGMENTS, LINES, AND ANGLES

You should know these words that are used often in geometry.

LINES AND SEGMENTS

Parallel

Two lines that do not meet are **parallel**.

Intersect

If two lines or segments meet, they **intersect** at a point.

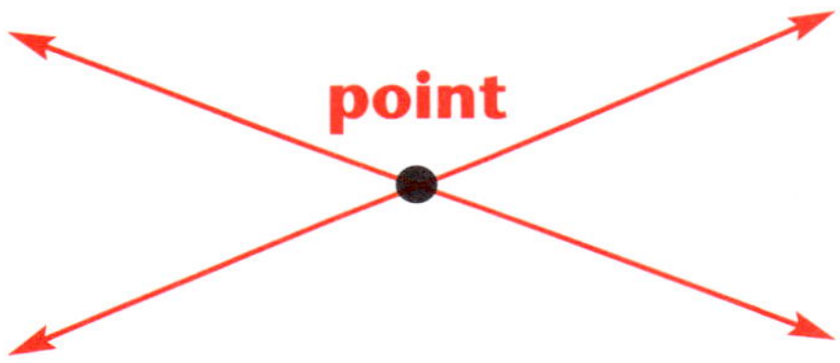

Perpendicular

If two lines or segments intersect to make square corners (right angles), they are **perpendicular**.

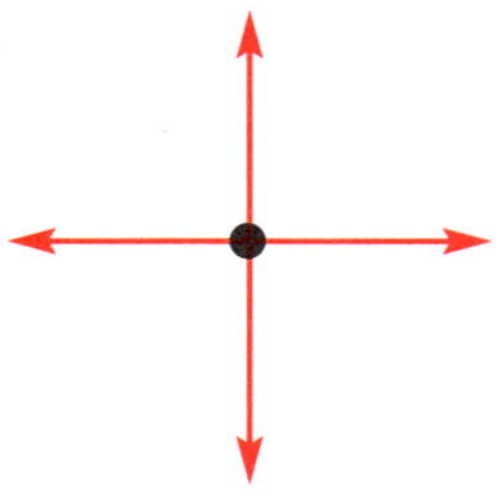

Midpoint of a segment

The midpoint of a segment divides the segment into two equal parts.

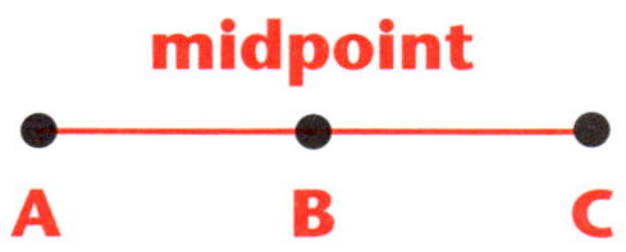

Length of $\overline{AB}$ = Length of $\overline{BC}$.
B is the midpoint of $\overline{AC}$.

ANGLES

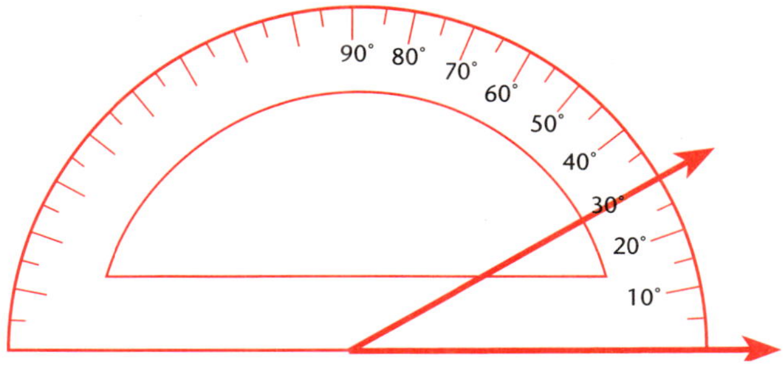

We measure angles with a protractor, like the one above.

The angle above measures **30 degrees (30°)**.

Right angle

An angle that measures 90° (a square angle).

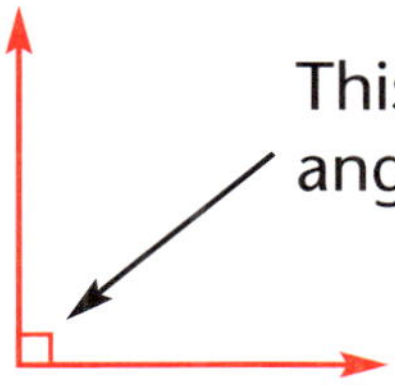

This symbol means that the angle is a right angle.

Acute angle

An angle that measures less than 90°.

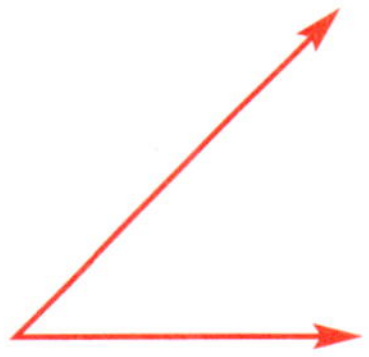

Obtuse angle

An angle that measures more than 90°.

Straight angle

An angle that measures 180°.

Sum of the angles of a triangle.

If you add the three angles of any triangle, the sum is 180°.

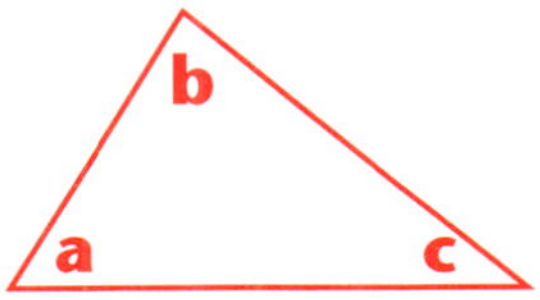

a + b + c = 180°

EXAMPLE

What is the measure of angle C?

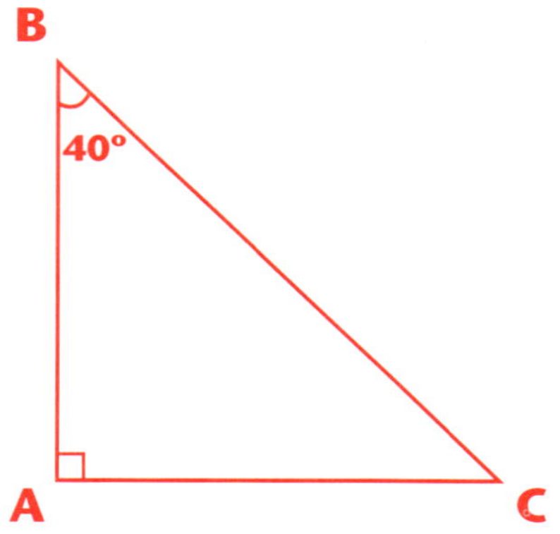

STRATEGY: **Add the angles of the triangle whose measures you know. Then subtract from 180 (the sum of the angles of a triangle).**

STEP 1: What are the measures of angles A and B?

Angle A measures 90°. It is a right angle.

Angle B measures 40°. It is an acute angle.

STEP 2: Add angles A and B.

Angle A	=	90°
Angle B	=	40°
		130°

STEP 3: Subtract 130° from 180°.

$180° - 130° = 50°$

SOLUTION: **The measure of angle C is 50°.**

SAMPLE TEST QUESTIONS

1 Which angle is closest to 30°?

A

B

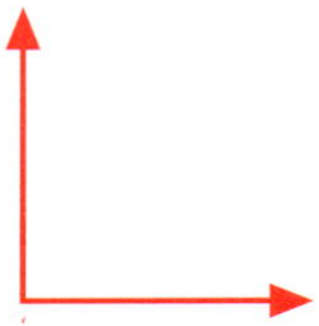

C

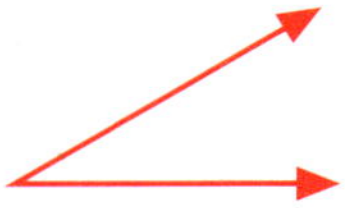

D

2 Which of these lines are perpendicular?

A

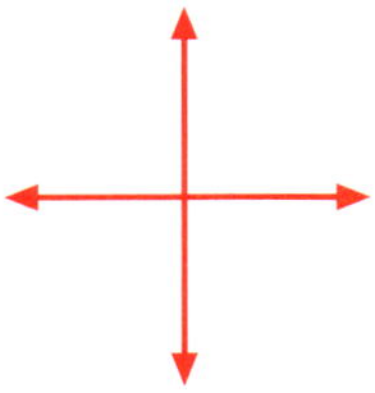

B

C

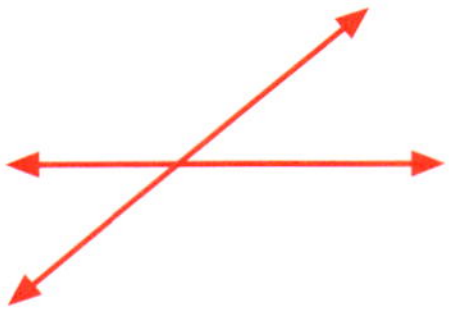

D

3 Which kind of angle is this?

A right angle

B acute angle

C obtuse angle

D straight angle

4 Points P, Q, and R are on the same line. The length from P to Q is the same as the length from Q to R. What is Q called?

A midpoint

B endpoint

C segment

D straight line

5 What is the measure of angle Z?

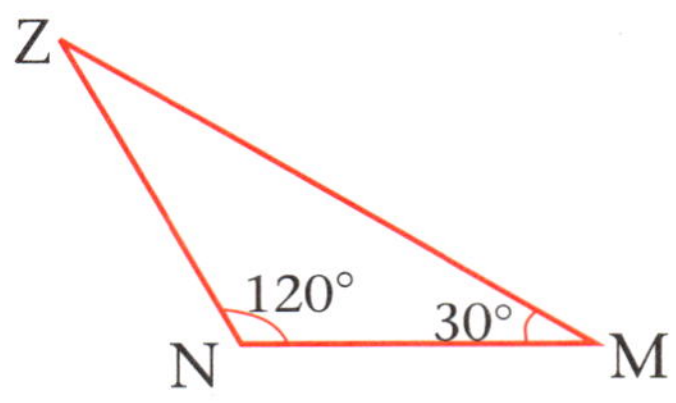

A 90° **B** 50°

C 40° **D** 30°

LESSON 9

LOCATING POINTS ON A GRID

Every point on a grid has an address. The address is the way to find the point.

EXAMPLE 1

What object is at (D,5)?

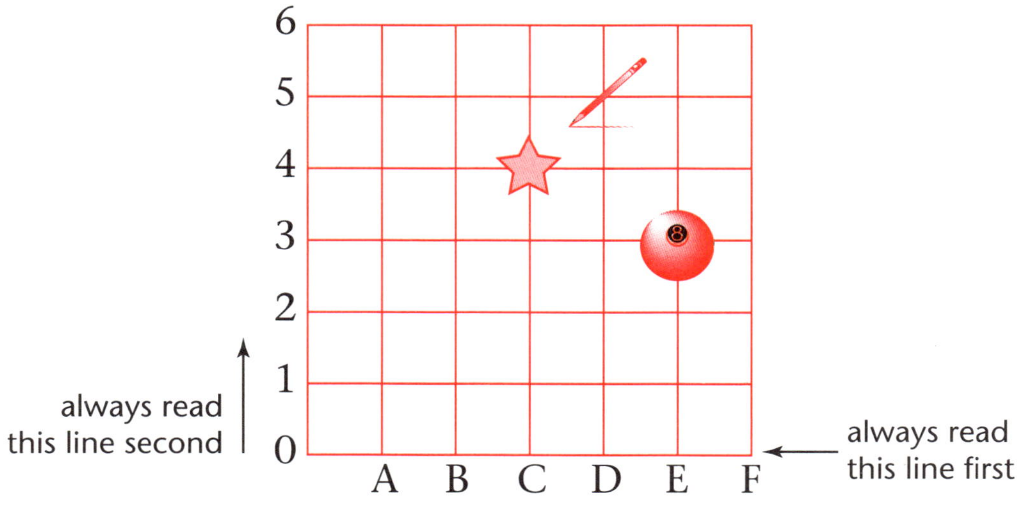

STRATEGY: **Use the two parts of the address (D,5) to find the object.**

STEP 1: Start at 0 on the grid. First, find the line with the letters.

Place your finger at D.

STEP 2: Now use the second part of the address—the number.

Move your finger up 5 spaces.

SOLUTION: **The object is a pencil.**

EXAMPLE 2

What pair of numbers is the address for the star?

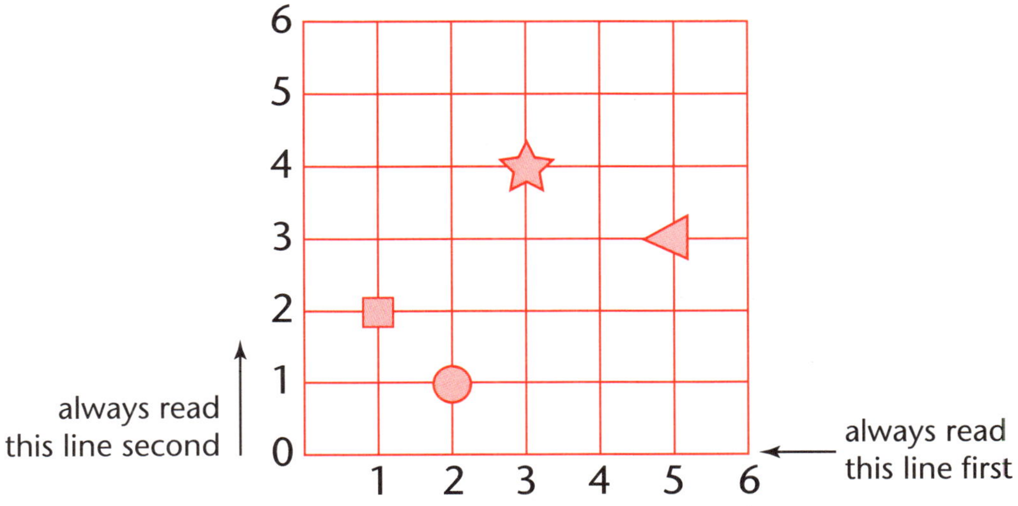

STRATEGY: **Use the two number lines to write an address for the star.**

STEP 1: Place your finger on the star.

STEP 2: Bring your finger down to the number line below the star.

Remember, you always start with the number line that goes left to right (not up and down).

What number is below the star?

The number is 3. This is the first number of the address.

STEP 3: Place your finger on the star again. Move your finger to the number line on the left.

The number is 4. This is the second number of the address.

STEP 4: Put these two numbers together.

SOLUTION: **The address is (3,4).**

EXAMPLE 3

Start at X. Go 2 blocks south, then go 4 blocks west. Where do you end up?

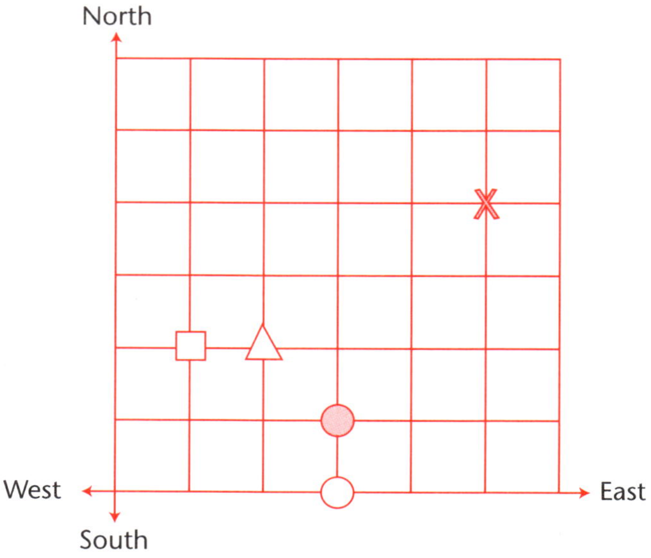

A □ **C** ●

B △ **D** ○

STRATEGIES: **Make sure you understand the 4 directions: north, south, east, west. Follow directions carefully.**

STEP 1: Place your finger on X.

This is the starting point.

STEP 2: Go 2 blocks south.

South is down. Count 2 spaces down. Keep your finger on this place.

STEP 3: Go 4 blocks west.

Count 4 blocks to the left. Keep your finger on this place.

STEP 4: Identify the shape here.

SOLUTION: **The place where you end up is □, answer A.**

SAMPLE TEST QUESTIONS

1 Where is point Q on the grid?

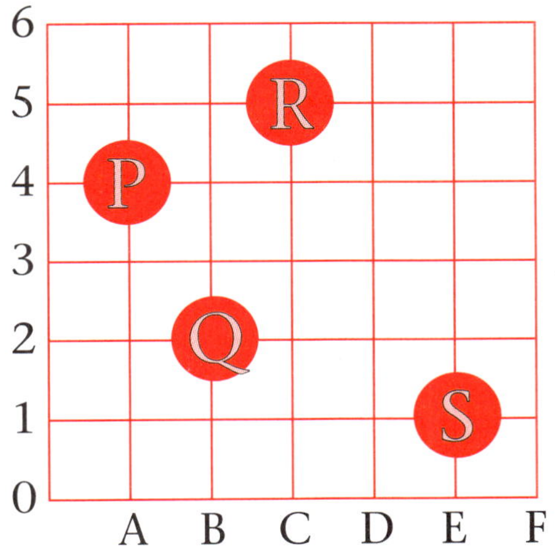

A (B,3) **B** (B,2)

C (B,5) **D** (C,3)

2 What object is located at (C,5)?

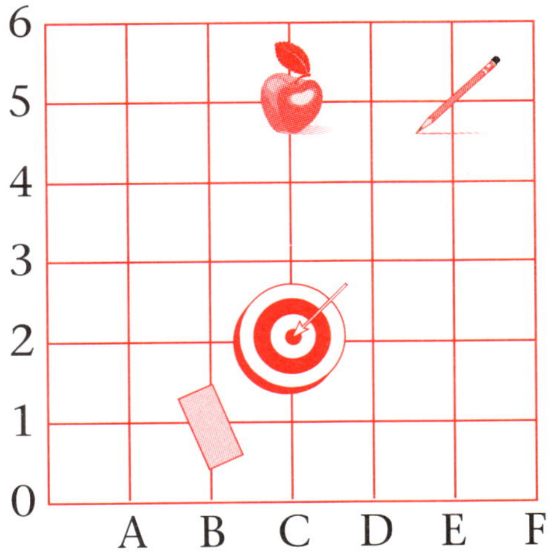

A apple

B bullseye

C pencil

D rectangle

3 What is the number pair, or address, for the flower?

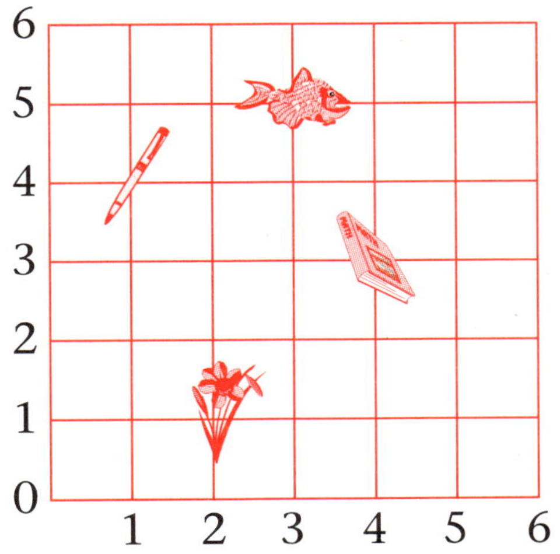

A (2,1) **B** (1,2)

C (2,4) **D** (2,4)

4 Start at ◁, go 2 blocks south. Then go 4 blocks east. Where do you end up?

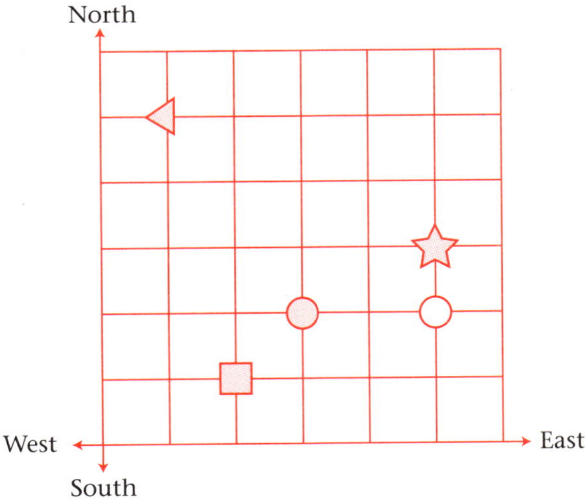

A  **B** 

C  **D** 

LESSON 10

ESTIMATING LENGTH

Here are some examples of lengths using customary and metric units.

CUSTOMARY UNITS

inch This segment is 1 inch long. ______

foot This page, from top to bottom, is a little less than 1 foot long.

1 foot (ft) = 12 inches (in.)

yard 1 yard (yd) = 3 feet (ft)

mile When you travel by car, airplane, or train, you measure distance in miles.

1 mile (mi) = 5,280 feet (ft)

METRIC UNITS

millimeter about this long -

centimeter about this long ___

1 centimeter (cm) = 10 millimeters (mm)

meter A meter is a little longer than 3 feet (1 yard).

1 meter (m) = 100 centimeters (cm)

kilometer 1 kilometer (km) = 1,000 meters (m)

EXAMPLE 1

What would be the best measure for the height of a door?

A 2 centimeters

B 10 centimeters

C 2 meters

D 5 meters

STRATEGY: **Find the correct answer by getting rid of answers that do not make sense.**

STEP 1: Think about Answer A:

2 centimeters is about this long:

Too small.

STEP 2: Think about Answer B:

10 centimeters is about this long.

Still too small.

STEP 3: Think about Answer C:

A meter is a little over 3 feet.

2 meters is therefore a little more than 6 feet.

That seems about right. But check Answer D anyway.

STEP 4: Think about Answer D:

A meter is over 3 feet long, so 5 meters is over 15 feet!

Way too big.

SOLUTION: **The correct answer is C, 2 meters.**

EXAMPLE 2

Juan is 5 feet 4 inches tall. Steve is 4 feet 11 inches tall. How much taller is Juan than Steve?

STRATEGY: **Find the difference between Juan's height and Steve's height.**

STEP 1: How many inches is it from 4 feet 11 inches to 5 feet?

There is 1 inch from 4 feet 11 inches to 5 feet.

STEP 2: How many inches is it from 5 feet to 5 feet 4 inches?

There are 4 inches.

STEP 3: Add the two lengths together.

1 inch + 4 inches = 5 inches.

SOLUTION: **Juan is 5 inches taller than Steve.**

SAMPLE TEST QUESTIONS

1 Which of these is probably about 10 feet high?

A a drinking fountain

B a fourth-grader

C the distance from the floor to the ceiling

D a school building

2 Which is the best unit to measure the length of a pen?

A mile

B centimeter

C meter

D kilometer

3 How much longer is segment MN than PQ?

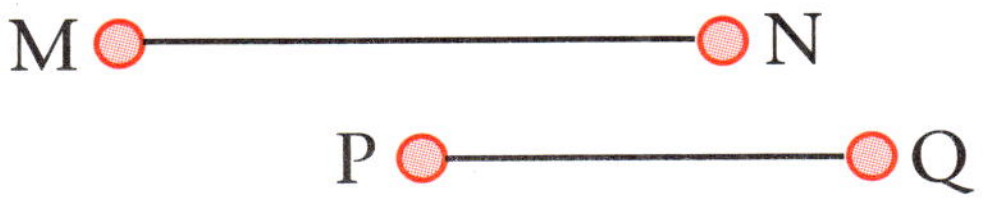

A 1 cm

B 1.5 cm

C 2 cm

D 2.5 cm

4 The width of a picture frame is 20 units. What unit was most likely used to measure the width?

A miles

B yards

C feet

D inches

5 A foot is about equal to how many centimeters?

A 1

B 10

C 30

D 100

6 How many meters are in 400 centimeters?

A 2

B 4

C 6

D 8

7 About how many miles is the same as 26,000 feet?

A 5

B 4

C 3

D 2

LESSON 11

ESTIMATING WEIGHT

Here are some weights using customary and metric units.

CUSTOMARY UNITS

ounce the weight of 5 grapes

pound the weight of a thick paperback book

1 pound (lb) = 16 ounces (oz)

ton 1 ton (T) = 2,000 pounds (lb)

A car weighs about 1 ton.

METRIC UNITS

gram about the weight of a raisin

kilogram a little more than 2 pounds

1 kilogram (kg) = 1,000 grams (g)

EXAMPLE

What would be the best measure of the weight of a textbook?

A 1 ounce

B 2 pounds

C 5 ounces

D 10 pounds

STRATEGY: **Get rid of answers that do not make sense.**

STEP 1: Think about Answer A.

1 ounce is the weight of about 5 grapes.

Too small.

STEP 2: Think about Answer B.

2 pounds is the weight of about 2 paperback books.

This seems about right.

But check Answers C and D anyway.

STEP 3: Think about Answer C.

5 ounces is the weight of about 25 grapes ($5 \times 5 = 25$).

Too small.

STEP 4: Think about Answer D.

10 pounds is the weight of about 10 paperback books.

Too big.

SOLUTION: **The correct answer is B, 2 pounds.**

SAMPLE TEST QUESTIONS

1 What measure would be the weight of a small dog?

A 7 g

B 70 g

C 7 kg

D 70 kg

2 What measure would be the weight of a truck?

A 5 pounds

B 50 pounds

C 5 tons

D 50 tons

3 A scale shows a weight of 10 grams. What object is most likely being weighed?

A a book

B a pencil

C a sandwich

D a carton of milk

4 About how much does this weigh?

A 5 grams

B 50 grams

C 500 grams

D 5,000 grams

5 Felicia is in the 4th grade. She weighs about 30 ______.

A kilograms

B grams

C pounds

D ounces

LESSON 12

ESTIMATING CAPACITY

Here are some measures you should know. Capacity means how much liquid a container, like a bottle, can hold.

CUSTOMARY UNITS

teaspoon a small spoon used to mix cocoa

tablespoon a large spoon used for soup (about 3 teaspoons)

1 tablespoon = 3 teaspoons

cup the size of an ordinary cup, like a cup of cocoa

pint think of a pint of milk or juice

1 pint (pt) = 2 cups (c)

quart think of a quart of milk

1 quart (qt) = 2 pints (pt)

gallon think of a large container of milk

1 gallon (gal) = 4 quarts (qt)

METRIC UNITS

milliliter think of a drop of water

liter a little more than a quart

1 liter (L) = 1,000 milliliters (mL)

EXAMPLE 1

What unit of measure would be best for measuring the amount of water in a bathtub?

A cup

B pint

C quart

D gallon

STRATEGY: **Check each answer. Get rid of the answers that do not make sense.**

STEP 1: Think about Answer A.

A cup is too small.

STEP 2: Think about Answer B.

A pint is too small.

STEP 3: Think about Answer C.

A quart is also too small.

STEP 4: Think about Answer D.

A gallon is the largest measure.

SOLUTION: **The correct answer is C. The gallon is the best unit to use to measure the amount of water in a bathtub.**

EXAMPLE 2

If you put two containers of milk—each 1 pint—together, how much milk do you have?

A 2 cups

B 1 quart

C 1 liter

D 1 gallon

STRATEGY: **Use the information at the beginning of the lesson.**

STEP 1: Identify the correct units.

A pint is a customary unit. Therefore it is best to use customary units. That leaves out "liter" as an answer—only come back to it if none of the other answers makes sense.

STEP 2: What is equal to 2 pints?

1 quart = 2 pints

SOLUTION: **2 pints of milk is equal to 1 quart, Answer B.**

SAMPLE TEST QUESTIONS

1 Which of these is usually sold by the liter?

A soda

B bread

C potatoes

D popcorn

2 Which of these is the greatest total liquid capacity?

A 4 quarts

B 9 pints

C 8 cups

D 1 gallon

3 Daniel uses a measuring cup when he cooks. How much water is in the cup below?

A 2 cups **B** $2\frac{1}{2}$ cups

C 3 cups **D** $3\frac{1}{2}$ cups

4 Which is the best estimate for the capacity of a glass of milk?

A 2 milliliters

B 20 milliliters

C 200 milliliters

D 2,000 milliliters

5 About how much oil is in this container?

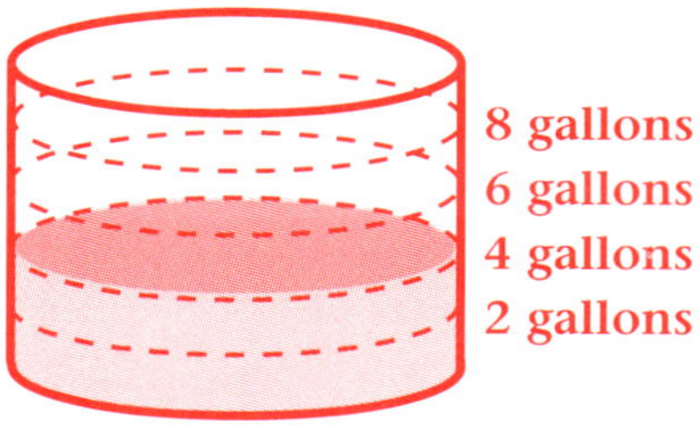

A about 8 gallons

B about 5 gallons

C about 3 gallons

D about 4 gallons

6 Manny drinks 6 glasses of water each day. He estimates that each glass holds 250 mL. About how many liters of water does he drink each day?

A about 2.5 L

B about 2 L

C about 1.5 L

D about 1 L

LESSON 13

COMPARING UNITS OF LENGTH, CAPACITY, AND WEIGHT

EXAMPLE 1

José's bedroom is 6 meters long. How many centimeters long is his room?

STRATEGY: **Use the information about centimeters and meters in Lesson 10.**

STEP 1: Find the number of centimeters in 1 meter.

From Lesson 10, you know that 1 m = 100 cm.

STEP 2: Do the math:

1 m = 100 cm   6 m = 600 cm

SOLUTION: **José's room is 600 cm long.**

EXAMPLE 2

Which is longer, 2 inches or 3 centimeters?

STRATEGY: **Use the information from Lesson 10.**

STEP 1: Compare the length of the lines for 1 inch and 1 centimeter.

1 inch is more than twice as long as 2 centimeters.

1 centimeter < $\frac{1}{2}$ inch

STEP 2: Do the math.

1 centimeter < $\frac{1}{2}$ inch

2 centimeters < 1 inch

3 centimeters < $1\frac{1}{2}$ inches

(and therefore < 2 inches)

SOLUTION: **2 inches is longer than 3 centimeters.**

EXAMPLE 3

Stella sent 3 packages by air freight for overnight delivery. The packages weighed 3 lb 12 oz, 4 lb 5 oz, and 2 lb 9 oz. What is the total weight in ounces of the three packages?

STRATEGY: **Use the information about pounds and ounces in Lesson 11.**

STEP 1: Add the three weights together.

	3 lb 12 oz
	4 lb 5 oz
+	2 lb 9 oz
	9 lb 26 oz

STEP 2: Use the information in Lesson 11 to regroup pounds to ounces.

9 lb = 9 × 16 oz = 144 oz

STEP 3: Add the ounces together.

9 lb 26 oz = 144 oz + 26 oz = 170 oz

SOLUTION: **The total weight of the three packages is 170 ounces.**

SAMPLE TEST QUESTIONS

1 Don's fishing pole measures 1 yd 1 ft 4 in. How many inches long is the pole?

A 50 in.

B 52 in.

C 56 in.

D 62 in.

2 If a liter bottle is half-filled with juice, how much more juice is needed to fill the bottle?

A 200 mL

B 300 mL

C 400 mL

D 500 mL

3 Which is the longest?

A a 20 cm flashlight

B a 3 km race

C a 12 m sidewalk

D a 300 mm wire

4 Tara's mother's car weighs 1 T 400 lb. How many pounds does it weigh?

A 1,400 lb

B 2,200 lb

C 2,400 lb

D 2,600 lb

5 Gareth was making cookies, but he lost his measuring cup. All he has to measure with is a 1 quart container of milk, which he emptied out and cleaned. If the recipe calls for 2 cups of flour, how far should he fill the quart container?

A $\frac{1}{8}$ full

B $\frac{1}{4}$ full

C $\frac{1}{2}$ full

D full

LESSON 14

ELAPSED TIME

Here is what you should know about time and calendars.

1 **minute** = 60 seconds

1 **hour** = 60 minutes

1 **day** = 24 hours

1 **week** = 7 days

1 **month** = 4 weeks

1 **year** = 12 months

EXAMPLE 1

John is leaving on a vacation in 12 days. Today is Thursday. On what day of the week is John leaving for vacation?

STRATEGY: **Remember, 1 week equals 7 days.**

STEP 1: Change 12 days to weeks and days.

12 days = 7 days + 5 days = 1 week + 5 days

STEP 2: Count 5 days from Thursday. (A week from Thursday is next Thursday.)

Start at Thursday: Friday, Saturday, Sunday, Monday, Tuesday.

SOLUTION: **John will leave for vacation on Tuesday.**

EXAMPLE 2

How much time has elapsed since 3:00?

STRATEGY: **Find the difference between the two times.**

STEP 1: Find the time on the clock.

The time is 5:15, or 15 minutes after 5.

STEP 2: How many hours from 3 o'clock to 5 o'clock?

5 – 3 = 2 hours

STEP 3: How many minutes from 5 o'clock to 5:15?

It is 15 minutes from 5 to 5:15.

SOLUTION: **The time elapsed from 3 o'clock to 5:15 is 2 hours 15 minutes.**

SAMPLE TEST QUESTIONS

1 How much time has passed since 9:30?

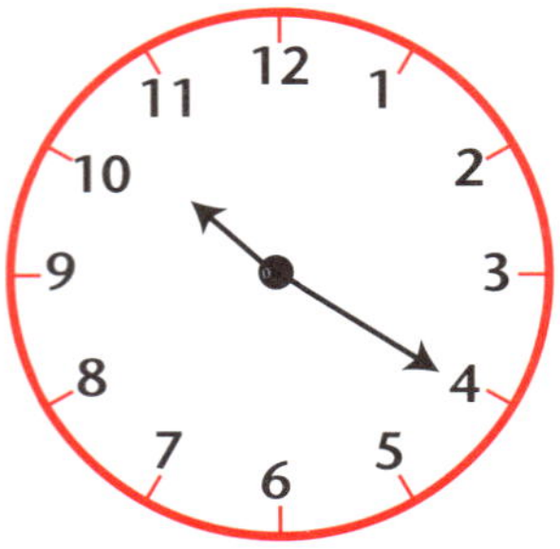

A 30 minutes

B 40 minutes

C 50 minutes

D 1 hour

2 Mrs. Brown spends 3 hours each Wednesday fixing computers. How many hours did she fix computers in September?

SEPTEMBER

Sun	Mon	Tues	Wed	Thurs	Fri	Sat
	1	2	3	4	5	6
7	8	9	10	11	12	13
14	15	16	17	18	19	20
21	22	23	24	25	26	27
28	29	30				

A 6 hours

B 9 hours

C 12 hours

D 15 hours

3 What is another way of expressing 74 minutes?

A 1 hour 24 minutes

B 1 hour 14 minutes

C 1 hour 12 minutes

D 1 hour 8 minutes

4 What is the fifth month of the year?

A April

B March

C June

D May

5 Felix's birthday is on March 12. Donna's birthday is 2 weeks later. When is Donna's birthday?

A March 26

B March 25

C March 23

D March 22

LESSON 15

MONEY

You should know this information about money.

1 **nickel**	=	5 pennies						
1 **dime**	=	2 nickels	=	10 pennies				
1 **quarter**	=	5 nickels	=	25 pennies				
1 **dollar**	=	4 quarters	=	10 dimes	=	20 nickels	=	100 pennies

EXAMPLE 1

How would you write two hundred ninety-four dollars and seventeen cents?

STRATEGY: **Say the amount slowly and write it as you say it to yourself.**

STEP 1: Remember to write a dollar sign ($) first.

STEP 2: Two hundred...

$**2**

STEP 3: ...ninety-four dollars...

$2**94**

STEP 4: ... and seventeen cents.

Write a decimal point (.) after 94, to show that you're writing cents now. Then write **17**.

SOLUTION: **$294.17.**

EXAMPLE 2

Which of these is the largest amount of money?

A

B

C

D

STRATEGY: **Find the value of each set of coins. It helps to change large groups of coins into dollars and coins. We call this regrouping. For example, 5 quarters = 1 dollar, 1 quarter.**

STEP 1: Regroup quarters.

In A: No regrouping

In B: 7 quarters = 1 dollar, 3 quarters

In C: 9 quarters = 2 dollars, 1 quarter

In D: 10 quarters = 2 dollars, 2 quarters

STEP 2: Use Step 1 to write the amount for each set of coins.

A: 4 dollars, 1 quarter, 3 pennies = $4.28

B: 3 dollars, 7 quarters, 3 pennies =
4 dollars, 3 quarters, 3 pennies = $4.78

C: 2 dollars, 9 quarters, 4 pennies =
4 dollars, 1 quarter, 4 pennies = $4.29

D: 1 dollar, 10 quarters, 5 pennies =
3 dollars, 2 quarters, 5 pennies = $3.55

SOLUTION: **The largest amount is $4.78—answer B.**

EXAMPLE 3

Mrs. Vann bought 4 scooters for her children. Each scooter cost $46.23. What was the cost of the 4 scooters?

STRATEGY: **Make sure you understand the problem.**

STEP 1: What operation do you need to solve this problem?

You use multiplication because you know the cost of one scooter and you want to know the cost of 4 scooters.

STEP 2: Do the math: $46.23 × 4 = $184.92

SOLUTION: **The cost of 4 scooters was $184.92.**

EXAMPLE 4

Jerry wanted to buy an action figure from the new *Space Battles* movie, so he broke open his piggy bank. He had 34 quarters, 3 dimes, 7 nickels, and 5 pennies, and the action figure cost $9.95. Could he afford it?

STRATEGY: **Regroup the quarters into dollars. Figure out how much he has and compare.**

STEP 1: If 4 quarters = 1 dollar, then 34 quarters = $8.50.

STEP 2: 3 dimes = $0.30, 7 nickels = $0.35,
5 pennies = $0.05.

STEP 3: $8.50 + $0.30 + $0.35 + $0.05 = $9.20

SOLUTION: **$9.20 is less than $9.95. No, Jerry could not afford the action figure.**

SAMPLE TEST QUESTIONS

1 Which of the following is the cost of the TV set?

A thirty-seven hundred and nine dollars

B three hundred seventy-nine dollars

C three dollars and seventy-nine cents

D thirty-seven dollars and nine cents

2 It cost $41.20 for lunch for 5 people. If the cost was divided equally, how much did each person pay?

A $7.24

B $8.28

C $6.25

D $8.24

3 Which is the smallest amount?

A 7 dollars, 2 quarters, 3 nickels

B 6 dollars, 8 quarters, 4 nickels

C 5 dollars, 9 quarters, 3 nickels

D 4 dollars, 14 quarters, 4 nickels

4 Kal had 7 quarters. He bought a toy for $1.62. How much change did he get?

A 8 cents

B 12 cents

C 13 cents

D 15 cents

5 Round this amount to the nearest dollar: 18 quarters, 15 dimes, 30 nickels.

A $6

B $7

C $8

D $9

LESSON 16

TEMPERATURE

Know these important temperatures.

	CUSTOMARY UNITS	METRIC UNITS
water freezes	32° Fahrenheit (32°F)	0° Celsius (0°C)
comfortable room temperature	68° Fahrenheit (68°F)	20° Celsius (20°C)
hot day	90° Fahrenheit (90°F)	32° Celsius (32°C)
water boils	212° Fahrenheit (212°F)	100° Celsius (100°C)

EXAMPLE

Which temperature is very hot?

A 0°C

B 10°C

C 22°C

D 34°C

STRATEGY: **Look at the chart above.**

SOLUTION: **Since 32°C is a hot day, 34°C is very hot.
The answer is D.**

SAMPLE TEST QUESTIONS

1 Kim is working in her office. She feels that the temperature is comfortable and that she does not need to wear a sweater. About what temperature would her office be?

A 20°F

B 30°F

C 68°F

D 90°F

2 Take a look at the thermometer. What kind of outdoor activity is this day good for?

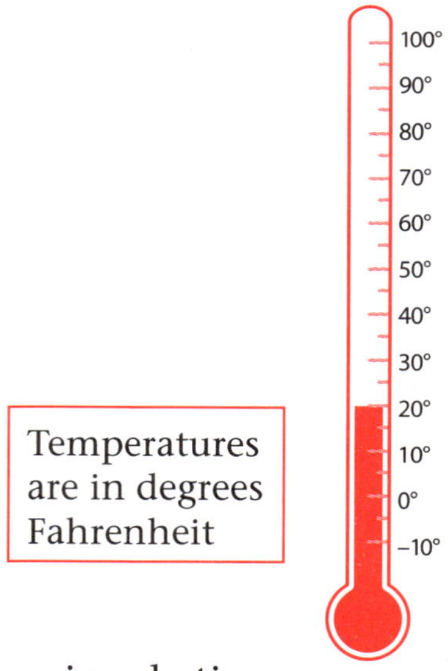

A ice-skating

B swimming

C a picnic

D jogging

3 Charlie checked the temperature in his room. It is 62°F. How many degrees above the freezing point is this?

A 20°F

B 25°F

C 30°F

D 35°F

4 What will the temperature be outside if it falls 15°C from what it shows now?

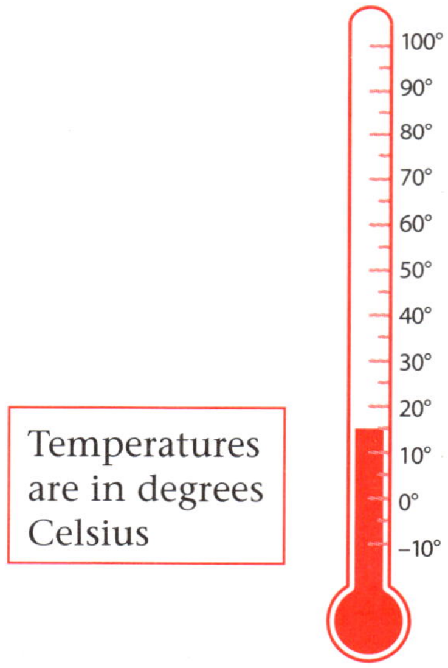

A 30°C

B 10°C

C 0°C

D 5° below 0°C

5 Harry said the temperature outside was at the freezing point. What temperature was it outside?

A 100°F

B 32°F

C 20°F

D 0°F

6 On any one day, what is the greatest difference between the high and low temperatures?

FOUR-DAY CITY CONDITIONS

	Wed	Thu	Fri	Sat
High	94°F	89°F	97°F	95°F
Low	73°F	67°F	78°F	74°F

A 19°F

B 21°F

C 22°F

D 24°F

7 The water from the faucet is very hot. What temperature could it be?

A 30°F

B 50°F

C 60°F

D 150°F

8 It was a great day to go swimming at the beach. What temperature could it be?

A 0°C

B 10°C

C 20°C

D 32°C

9 There was an unusual cold spell in April, and one day the temperature dropped 25°C from 22°C. What was the temperature after the drop?

A 3° below 0°C

B 3°C

C 30°C

D 47°C

LESSON 17

PERIMETER AND AREA

PERIMETER

The **perimeter** of a figure is the distance around the figure.

To find the perimeter of any polygon, add the lengths of the sides.

EXAMPLE 1

What is the perimeter of this rectangle?

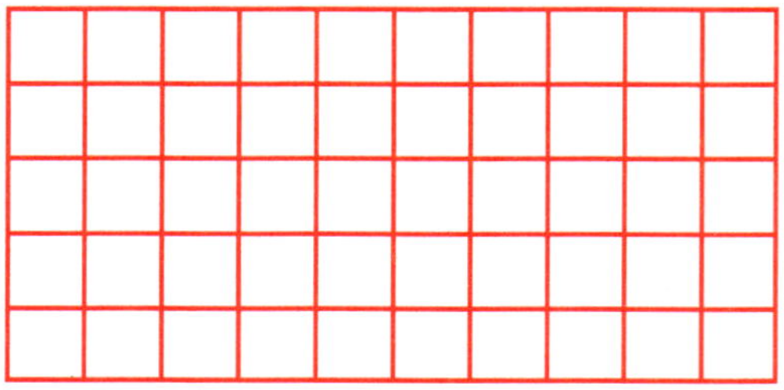

STRATEGY: **Count the number of squares along the sides of the rectangle.**

STEP 1: Count the number of squares along the two longer sides.

There are 10 squares on each of the two longer sides—that makes 20 units all together.

STEP 2: Count the number of squares along the two shorter sides.

There are 5 squares on each of the two shorter sides—that makes 10 units all together.

STEP 3: Add the number of units on the longer and shorter sides:

$$20 + 10 = 30 \text{ units}$$

SOLUTION: **The perimeter of the rectangle is 30 units.**

AREA

The area of a figure is the number of squares needed to cover it.

EXAMPLE 2

What is the area of the shaded triangle?

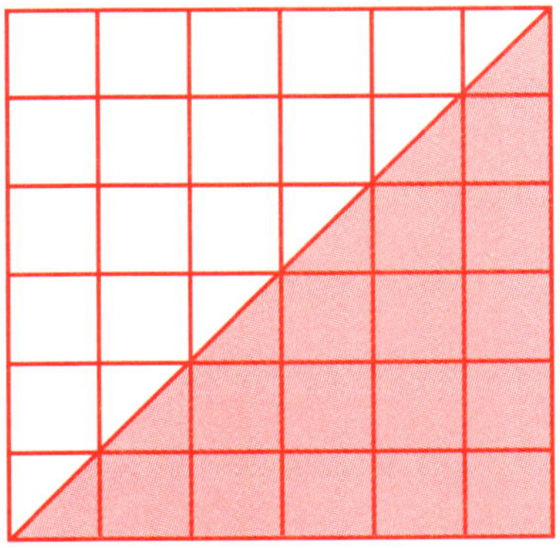

STRATEGY: **Start with the area of the square and take a fraction of it.**

STEP 1: What is the area of the square?

The area of the large square is the number of small squares that cover it.

Count the small squares. The area of the large square is 36 small squares.

STEP 2: What fraction of the square is the shaded triangle?

The shaded triangle is $\frac{1}{2}$ of the square.

STEP 3: Compute the area of the shaded triangle.

$\frac{1}{2}$ of $36 = 18$

SOLUTION: **The area of the shaded triangle is 18 square units.**

EXAMPLE 3

Can two rectangles each have a perimeter of 16 but different areas?

STRATEGY: **Draw different rectangles and test them.**

STEP 1: Pick a rectangle to test. A square is a good one to start with. (Remember, a square is a special type of rectangle.)

All four sides of a square are the same length.

If the perimeter is 16, each side is 16 ÷ 4 = 4.

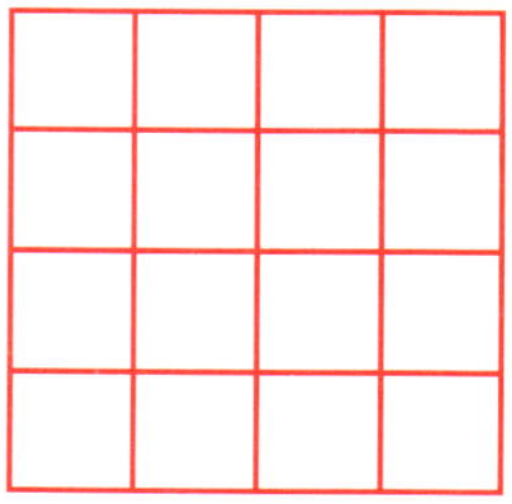

STEP 2: Use the figure of Step 1 to figure out the area.

The area is 16.

STEP 3: Try another rectangle.

Make sure the sides add up to 16.

Try it with long sides = 6 and short sides = 2.

(6 + 6 + 2 + 2 = 16)

STEP 4: Use the rectangle of Step 3 to figure out its area.

The area is 12.

SOLUTION: **Yes, two rectangles can have the same perimeters, but different areas.**

SAMPLE TEST QUESTIONS

1 A square measures 7.5 cm on each side. What is the perimeter?

A 15 cm
B 25 cm
C 30 cm
D 32 cm

2 What is the area of the shaded triangle?

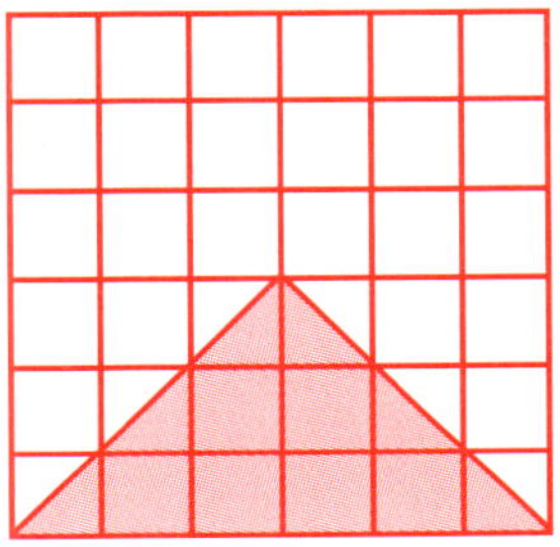

A 9 square units
B 10 square units
C 11 square units
D 12 square units

3 What is the perimeter of a rectangle that is 40 in. long and 22 in. wide?

A 62 in.
B 102 in.
C 124 in.
D 134 in.

4 What is the area of the shaded region?

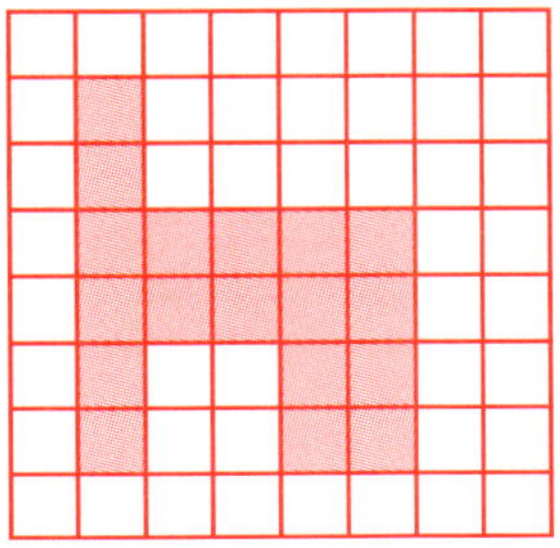

A 16 square units
B 18 square units
C 20 square units
D 22 square units

5 Which of the following has the same area as this figure?

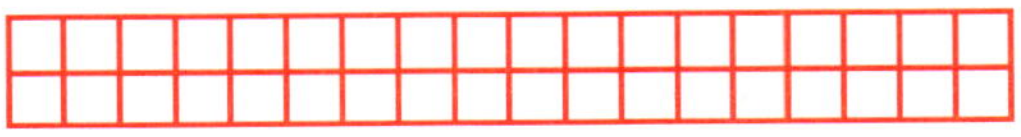

A

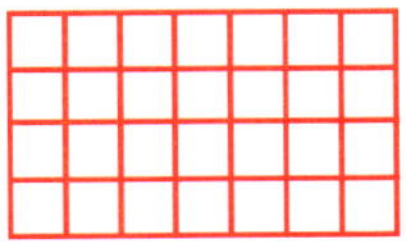

B

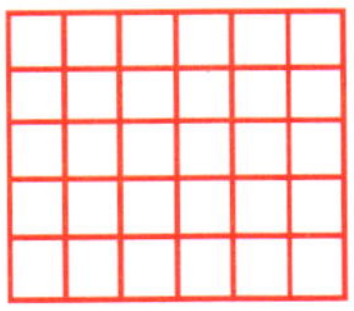

C

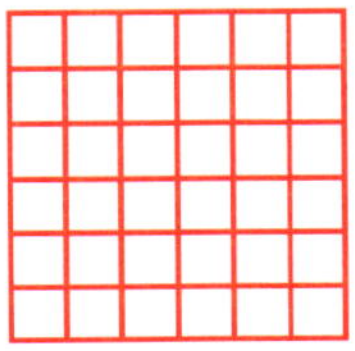

D

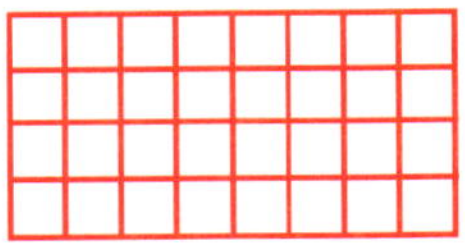

REVIEW OF LESSONS 5-17

1 Which name best describes these figures?

A rectangles

B squares

C polygons

D hexagons

2 What kind of sides does a rectangle have?

A parallel segments and no perpendicular segments

B perpendicular segments and no parallel segments

C no parallel segments and no perpendicular segments

D parallel segments and perpendicular segments

3 How many faces does a box have?

A 4

B 6

C 8

D 10

4 What kind of movement is shown in this figure?

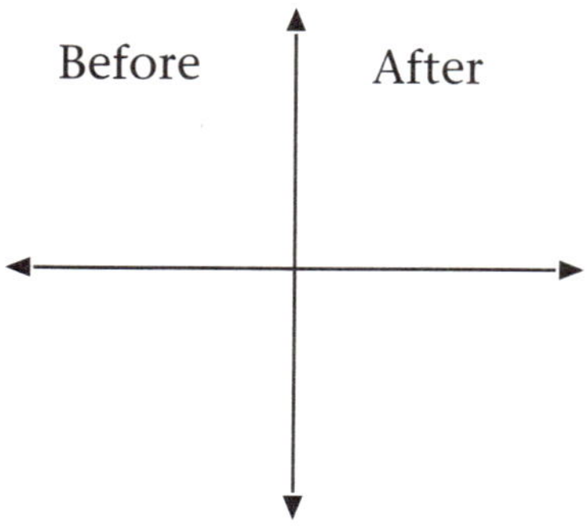

A turn

B slide

C flip

D jump

5 How many total right angles are there in this figure?

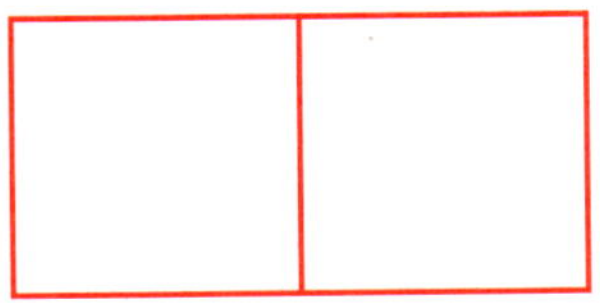

A 4

B 6

C 7

D 8

6 Which would be measured best by kilograms?

A a dozen potatoes

B a vitamin pill

C two paper clips

D a pencil

7 Which of these is usually sold by the quart?

A cookies

B orange juice

C sugar

D cereal

8 A gallon is the same as how many pints?

A 2

B 4

C 6

D 8

9 Jonathan's big school vacation comes two weeks after his birthday, which is on June 16. What is the date of his vacation?

A June 23

B June 25

C June 30

D July 2

10 Howard spent $6.42 on a toy and $7.21 on a book.
How much change did he get from $20?

A $7.37

B $6.37

C $6.27

D $6.17

11 What is the length of the third side of a triangle if two sides are each equal to 10 cm and the perimeter is 24 cm?

A 4 cm

B 6 cm

C 14 cm

D 20 cm

12 Shawn's summer vacation is 2 months long. About how many days long is his vacation?

A 20

B 40

C 60

D 80

13 Ashley said it was the hottest day of the summer. Which temperature is it most likely to be?

A 96° F

B 80° F

C 20° C

D 10° C

14 Which figure has the same area as the square?

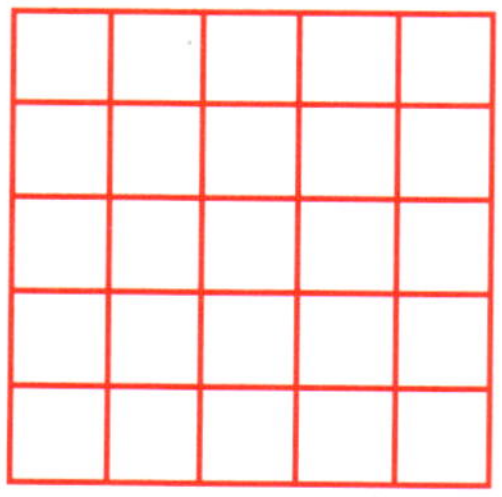

A

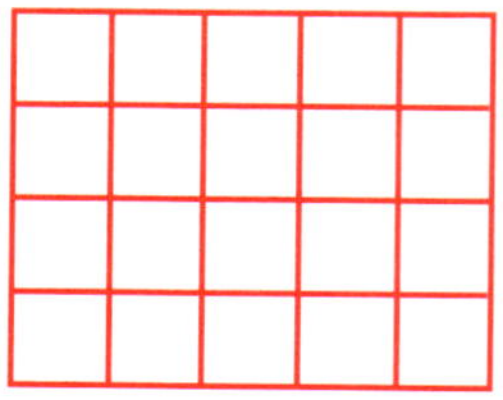

B

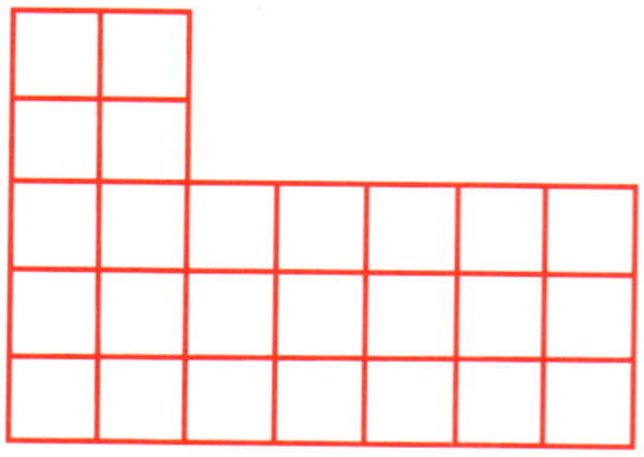

C

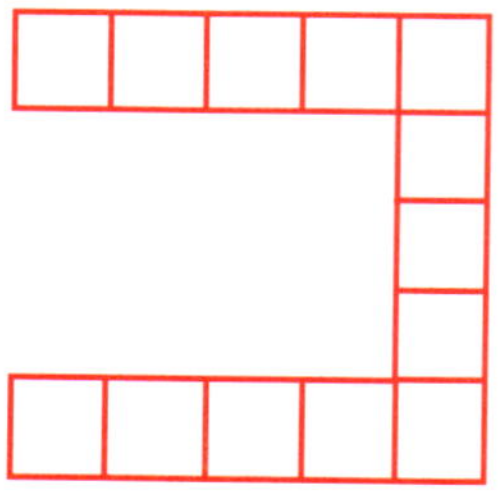

D

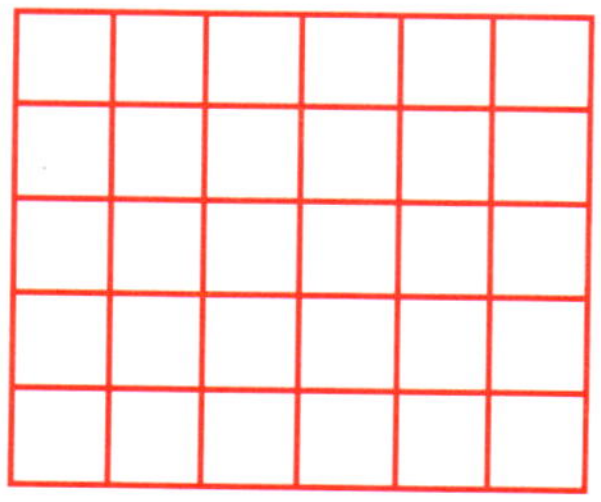

15 Every day, Caroline drives 5 miles to work. On the way home, she stops at a store that is 4 miles from her job. From the store to her house is 3 miles. What figure is her route, and what is its perimeter?

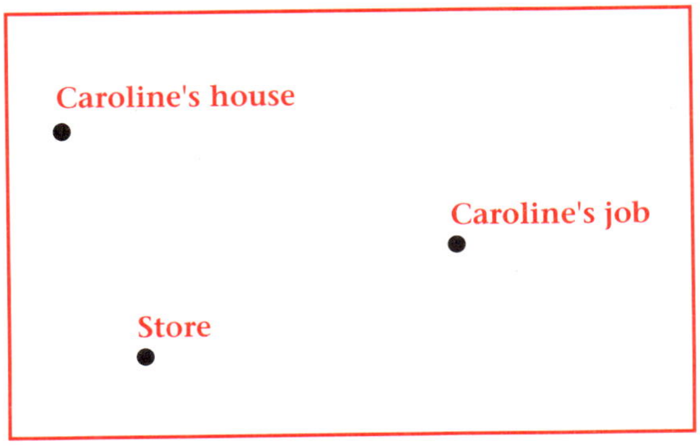

A A square. The perimeter is 12.

B A triangle. The perimeter is 10.

C A circle. The perimeter is 10.

D A triangle. The perimeter is 12.

16

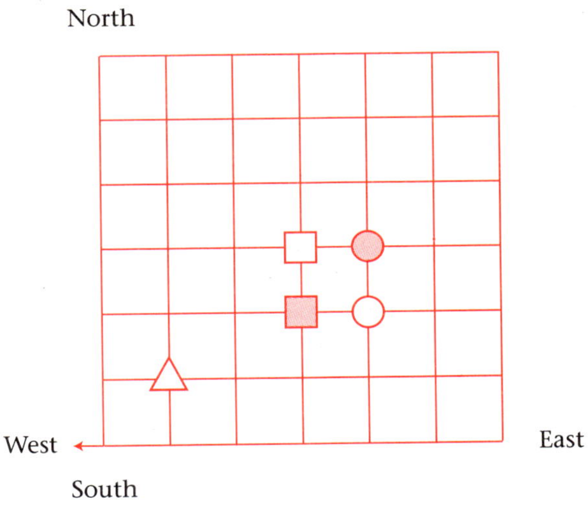

Start at △. First go 2 blocks north; then go 3 blocks east, and 1 block south. Where do you end up?

A 

B 

C 

D 

OPEN-ENDED QUESTIONS

1 Show how the letter A would look after a flip, a turn, and a slide:

a) flip over the line

b) half turn around point Z

c) slide 2 inches to the right

2 Draw an example for each of the following:

a) two parallel lines

b) two intersecting lines

c) one line perpendicular to a second line

d) an acute angle

e) a figure with five vertices

f) two congruent triangles

3 With your centimeter ruler, draw two rectangles with different lengths and widths that both have perimeters equal to 12 cm.

4 Joanna drinks 400 mL of juice each day. How many liters of juice does she drink after 2 weeks? Explain how you found your answer.

5 Draw two rectangles, each with an area of 12 square units, that have different perimeters.

LESSON 18

INTERPRETING INFORMATION FROM GRAPHS

A **graph** is a way to show and compare numerical information. There are four kinds of graphs you should know about.

PICTOGRAPHS

Pictographs are graphs that use pictures.

EXAMPLE 1

This pictograph shows you how many sunny days there were last spring. How many sunny days were there in June?

NUMBER OF SUNNY DAYS IN SPRING

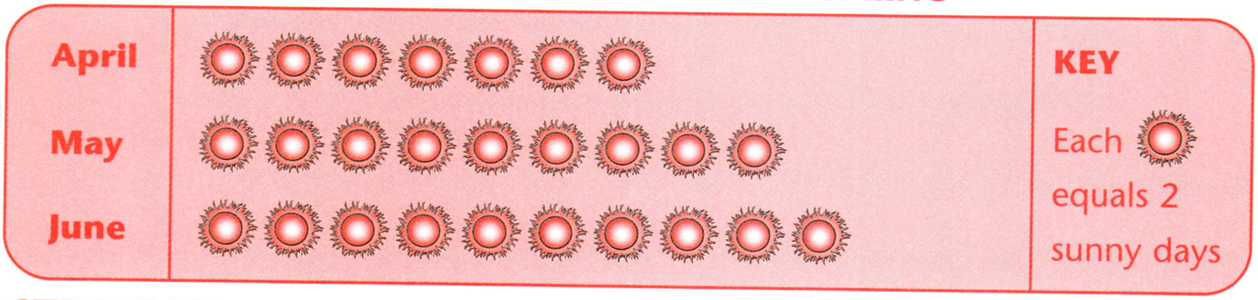

STRATEGY: **Count the suns in June. Then use the KEY.**

STEP 1: Find June on the graph. Then count the suns.

There are 10 suns.

STEP 2: Look at the KEY.

It tells you that each picture of the sun = 2 sunny days.

STEP 3: Find the total number of sunny days in June.

Multiply $10 \times 2 = 20$ sunny days.

SOLUTION: **There were 20 sunny days in June.**

BAR GRAPHS

Bar graphs use bars of different heights to compare things.

EXAMPLE 2

The graph shows how many books were read by four children in the fourth grade. Who read the most books, and how many did that person read?

NUMBER OF BOOKS READ BY 4 FOURTH–GRADERS

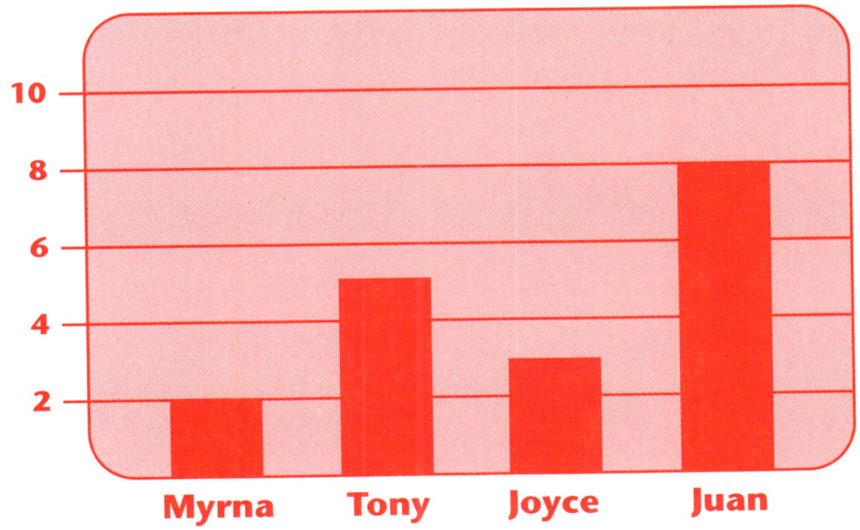

STRATEGY: **Find the highest bar. Then read the number on the scale that goes up and down.**

STEP 1: Look at all the bars.

The highest bar is the one for Juan.

STEP 2: From the top of the highest bar, follow the line to the number on the left.

It tells you the number of books Juan read.

STEP 3: The line for Juan's bar is 8.

SOLUTION: **Juan read the most, 8 books.**

CIRCLE GRAPHS

A **circle graph** divides a circle into parts. Each part represents a different category. When comparing two categories, there is more "stuff" in the category that takes up more of the circle.

EXAMPLE 3

Students in Mrs. Johnson's class were asked to name their favorite sport. She placed the results in this circle graph.

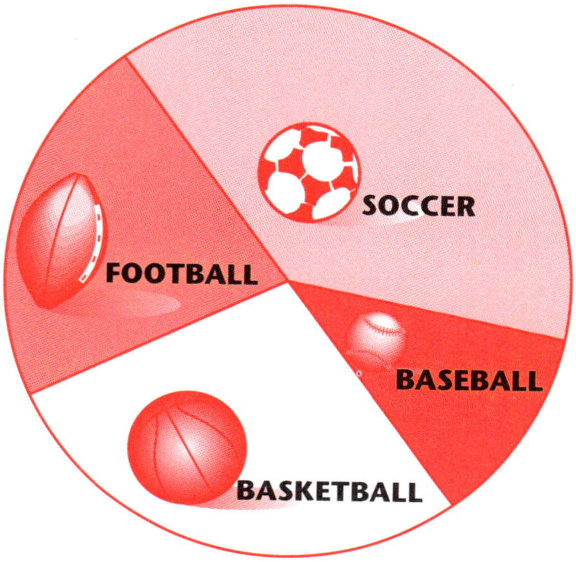

List the four sports in order from the most popular to the least popular.

STRATEGY: **Compare the areas of the four categories of the circle graph.**

STEP 1: Find the region with the greatest area.

The part that shows soccer has the greatest area.

STEP 2: Place the other sports in order according to the size of their areas.

basketball, football, and baseball

SOLUTION: **The most popular sport is soccer, then basketball, then football, and then baseball. Baseball is the least popular.**

LINE GRAPHS

In a line graph, a line shows a change or a trend.

EXAMPLE 4

The graph below shows the amount of TV watched over five weeks by a student.

NUMBER OF HOURS OF TV WATCHED OVER 5 WEEKS

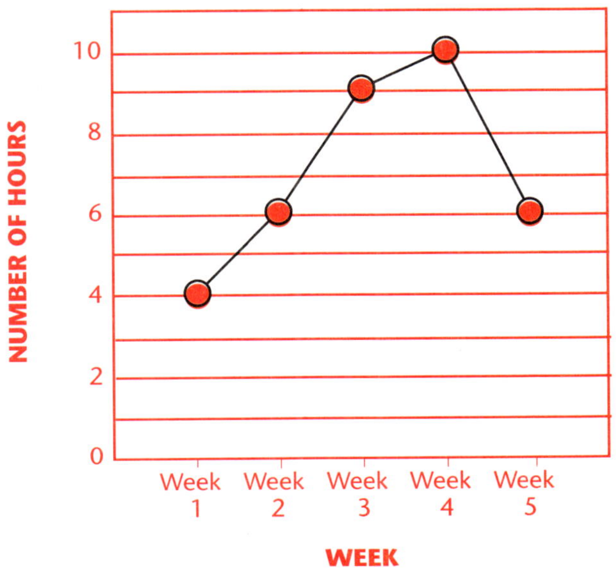

How many more hours of TV did the student watch during Week 3 than during Week 1?

STRATEGY: **Compare the number of hours for the two weeks and subtract.**

STEP 1: Find Week 1 and go up to the dot.

STEP 2: Move left from this dot to the scale opposite this dot and read the number.

The scale number is 4 hours.

STEP 3: Repeat Steps 1 and 2 for Week 3.

The scale number is 9 hours.

STEP 4: Subtract: 9 hours – 4 hours = 5 hours.

SOLUTION: **The student watched 5 more hours of TV on Week 3 than on Week 1.**

SAMPLE TEST QUESTIONS

1 What type of graph is shown here?

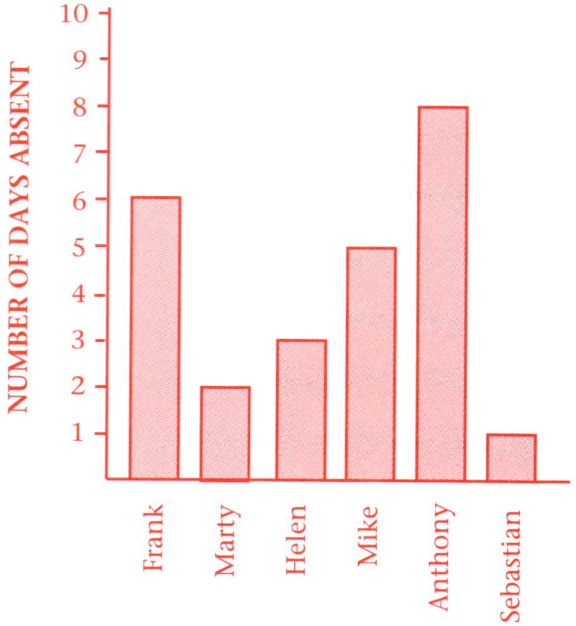

A line graph

B bar graph

C circle graph

D pictograph

2 This graph shows the value of a new car over the first five years after purchase.

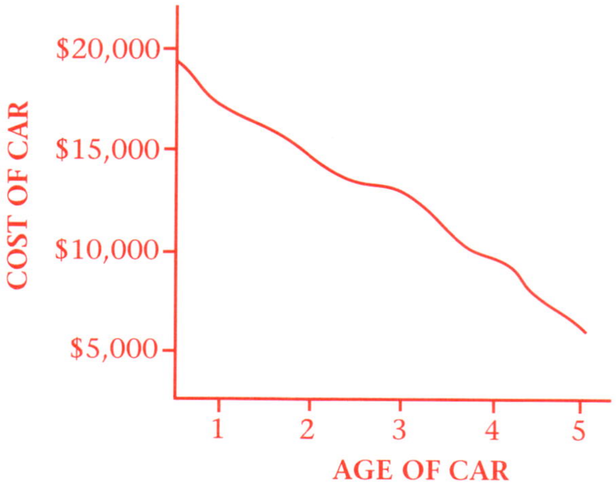

Which statement fits the graph?

A The value of a car increases in the first five years.

B The value of a car remains roughly the same for the first five years.

C The value of a car decreases in the first five years.

D The value of a car increases and decreases several times over the first five years.

3 How many students were enrolled in District 3?

STUDENTS ENROLLED IN EACH DISTRICT

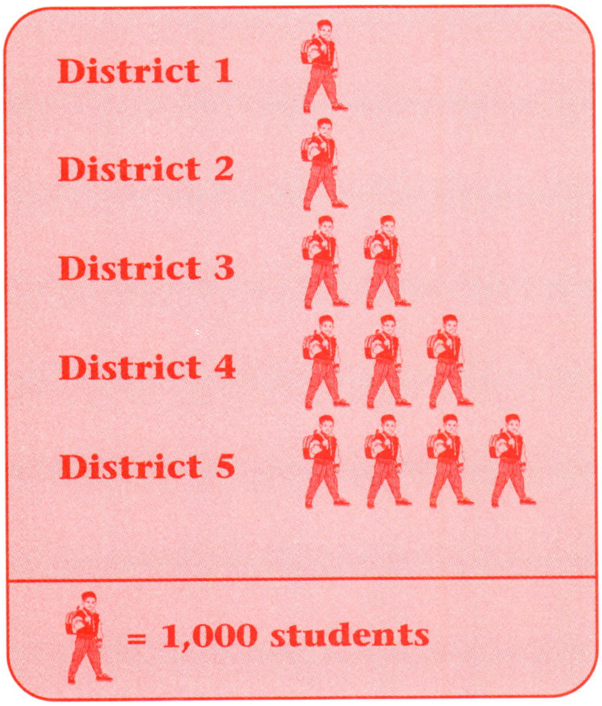

A 1,000

B 2,000

C 3,000

D 4,000

4 Last month, Darryl saw 10 movies, Bart saw 8 movies, Willie saw 7 movies, and Hadassah saw 6 movies. In the graph below, which bar shows the number of movies Willie saw?

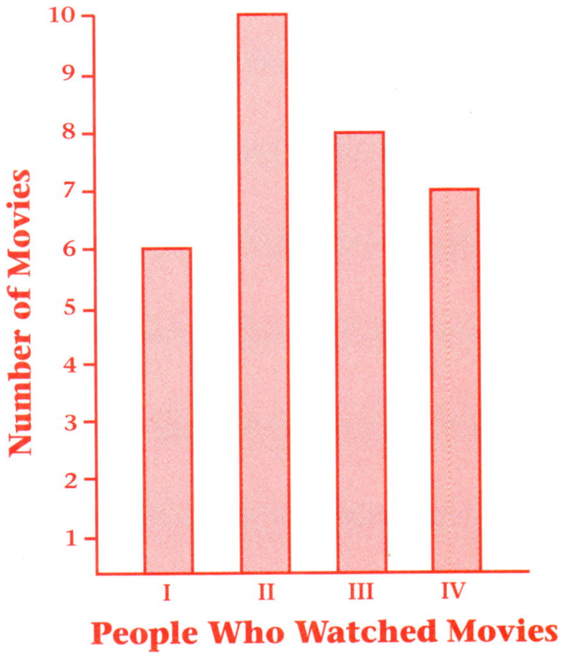

A I

B II

C III

D IV

5 Manny made a graph of how he spent his money last year.

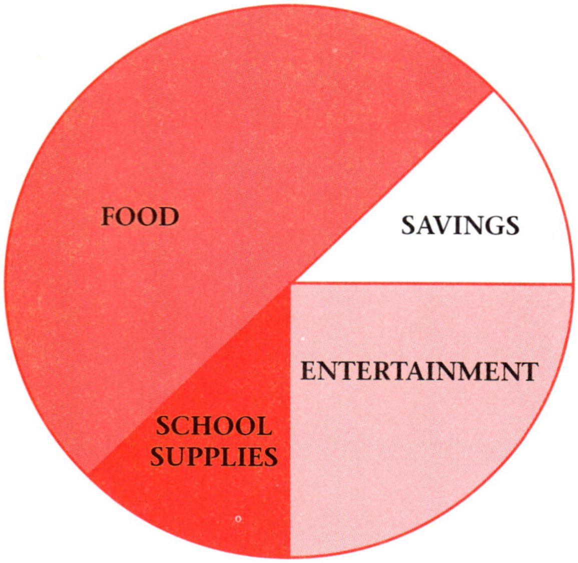

On what did he spend $\frac{1}{2}$ of his money?

A food

B savings

C entertainment

D school supplies

6 What is the name of a graph that uses pictures?

A pictograph

B line graph

C circle graph

D bar graph

7 You want to show how the number of students in your school increased over the past five years. What kind of graph would be the best way to show this trend?

A bar graph

B pictograph

C line graph

D circle graph

8 What kinds of graphs are good ways to compare quantities?

A segment and bar graphs

B circle and line graphs

C point and bar graphs

D circle and bar graphs

LESSON 19

INTERPRETING INFORMATION FROM CHARTS

CHARTS

A chart or table shows information in an organized way. Making a chart or table is a good way to collect information.

Some charts use tally marks, or tallies, to show information. The chart below uses tallies.

EXAMPLE 1

How many children walk to school each morning?

STRATEGY: **Just count the tallies in the correct row. Remember that 卌 stands for the number 5. Follow these steps:**

STEP 1: The question asks for the number of children who walk. Find the row in the table for walking.

The first row shows the number of children who walk to school. You don't need to look at the tallies in the other rows.

STEP 2: Count the tallies in the first row.

Remember, 卌 means 5 children, so start by counting by 5's.

5, 10, 15, 20, ..., then count by 1's: 21, 22, 23.

SOLUTION: **There are 23 children who walk to school.**

SAMPLE TEST QUESTIONS

1 **AMOUNT OF TV WATCHED BY FOURTH-GRADERS**

About $\frac{1}{2}$ Hour or Less	𝍸 |
About 1 Hour	𝍸 𝍸
About $1\frac{1}{2}$ Hours or more	𝍸 𝍸 |

This chart shows the amount of TV Mrs. Green's fourth-grade class watched last Wednesday night. How many students watched $1\frac{1}{2}$ hours or more?

A 6

B 9

C 11

D 16

2 There were 3,000 students at the school's basketball game. To find out their favorite foods, a poll was taken of 30 students.

FAVORITE FOODS AT GAME

FOOD	NUMBER OF STUDENTS
Hot Dogs	3
Hamburgers	13
Pizza	12
Submarine Sandwich	2

What is the best conclusion to make based on these results?

A More students at the game liked hamburgers than pizza.

B All students at the game liked one of the four foods.

C Hamburgers were the most popular food at the game.

D About the same number of students at the game liked hamburgers as pizza.

3 Scott made a chart of the lemonade sold at his stand last summer.

SIZE OF GLASS	NUMBER OF CUPS
Large	65
Medium	38
Small	86

Which question can be answered by this table?

A How many people bought lemonade from Scott's lemonade stand?

B How many cups did each person buy?

C How many people had more than one cup?

D None of the above.

4 This chart shows the ages of 24 fourth-grade students in Mr. Mohegan's class.

AGE	NUMBER
8	1
9	19
10	4

How many students are older than 9 years?

A 1

B 4

C 19

D 23

REVIEW OF LESSONS 18-19

1 This graph shows the amount of ice cream that students eat in the school cafeteria during winter and spring.

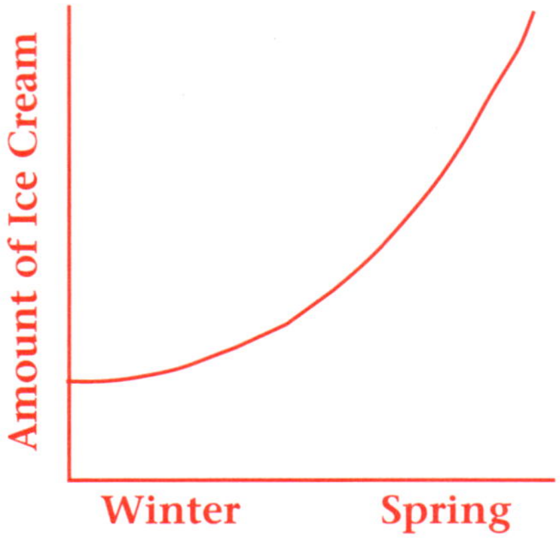

What does the graph show?

A that students eat more ice cream in the spring than in the winter

B that students eat about the same amount of ice cream in the winter as in the spring

C that students eat less ice cream in the spring than in the winter

D that students eat more ice cream in the winter than in the spring

2 **CARS PARKED BEHIND SCHOOL BUILDING**

MONDAY	22
TUESDAY	21
WEDNESDAY	25
THURSDAY	19
FRIDAY	17

This chart shows the number of cars parked in the school parking lot last week. How many more cars were parked there on Wednesday than on Monday?

A 4 **B** 3

C 2 **D** 1

3 This graph shows the number of students who were on Justin's bus last week.

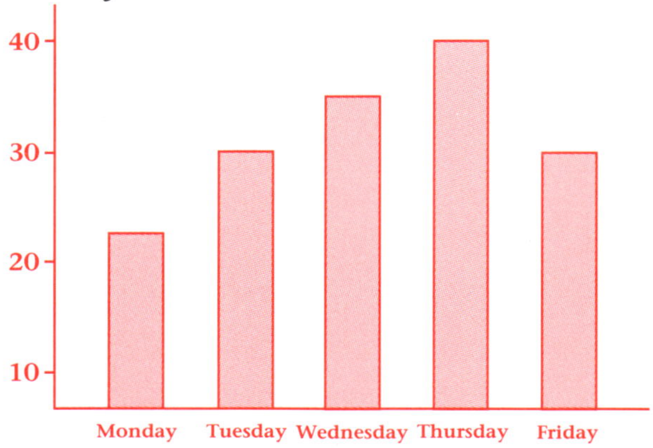

How many students were on Justin's bus on Wednesday?

A 30 **B** 35

C 40 **D** 45

OPEN-ENDED QUESTIONS

1 Complete the pictograph using the following table.

SOCCER BALLS USED IN THE TOURNAMENT

Game Number	Number of Soccer Balls
Game 1	12
Game 2	14
Game 3	10
Game 4	12
Game 5	11

SOCCER BALLS USED IN THE TOURNAMENT

Game Number	Number of Soccer Balls
Game 1	
Game 2	
Game 3	
Game 4	
Game 5	

Use ● = 2 soccer balls

2 Mike and Dana took a survey of 100 students in the fourth grade of the Burlington Elementary School to find out what their favorite sports are:

Football	—	**25**	**Soccer**	—	**10**
Basketball	—	**35**	**Other**	—	**30**

Display this data in a graph.

LESSON 20

PLACE VALUE AND WRITING NUMBERS

This is a **place value chart**. It shows the values of each digit for the number 824,517.

hundred thousands	ten thousands	thousands	hundreds	tens	ones
8	2	4	5	1	7

EXAMPLE 1

What is the value of **3** in 6**3**7,259?

STRATEGY: **Use a place value chart.**

STEP 1: Place the digits of 637,259 in the chart.

hundred thousands	ten thousands	thousands	hundreds	tens	ones
6	**3**	7	2	5	9

STEP 2: Read the value of **3** from the chart.

SOLUTION: **The value of 3 is 3 ten thousands.**

EXAMPLE 2

The radio said that there were eighty-five thousand seven hundred two people at a football game. Write this number in standard numerical form.

STRATEGY: **Read this number to yourself. As you read it, follow the steps below.**

STEP 1: There are eighty-five thousand. Stop reading the number and write 85.

85

STEP 2: Write a comma after this number, to show that 85 refers to thousands.

85,

STEP 3: Continue reading the number. After 85 thousand, you have seven hundred two. Write 702 after the comma.

85,**702**

SOLUTION: **The number in standard form is 85,702.**

SAMPLE TEST QUESTIONS

1 What is the value of the **9** in 603,**9**42?

A 90,000 **B** 9,000

C 900 **D** 90

2 In the number 7**0**6,423, the **0** is in which place?

A tens place

B hundreds place

C thousands place

D ten thousands place

3 What number does this tally chart show? (Remember, 卌 = 5)

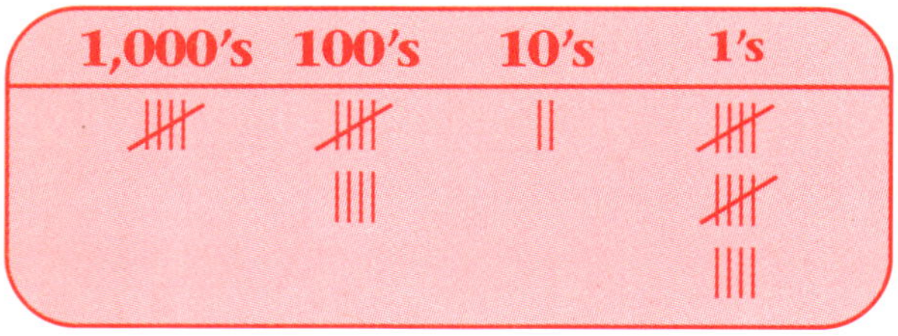

A 59,214 **B** 5,934

C 5,924 **D** 5,424

4 If ✳ = 1000, ✤ = 100, and @ = 10, what number is represented by the following?

✳✳✳✤✤✤✤✤✤@@

A 362,000 **B** 36,200

C 3,620 **D** 362

5 Felix saw a number in a book written this way:

7000 + 500 + 30 + 8

What is another way to write this number?

A 75.38

B 7,538

C 75,380

D 7,000,538

6 What digit is in the thousands place of 465,807?

A 4 **B** 6

C 5 **D** 8

7 Choose the correct way to say this number: 942,735.

A nine hundred forty-two thousand, seven hundred thirty-five

B nine hundred forty-two thousand, seven hundred fifty-three

C nine hundred forty-two million, seven hundred thirty-five

D ninety-four thousand, seven hundred thirty-five

LESSON 21

THE COMMUTATIVE PROPERTY AND THE PROPERTY OF ZERO

A rule is sometimes called a **property**. Here are a few properties that you will be using many times.

THE COMMUTATIVE PROPERTY—ADDITION

When you add numbers, the order in which you add them doesn't matter. The sum is still the same. This rule is called the **Commutative Property**.

Here are two ways to show the Commutative Property:

1. $7 + 5 = 12$

$5 + 7 = 12$

2. $7 + 5 = 5 + 7 = 12$

EXAMPLE 1

If $462 + 287 = 749$, then $287 + 462 = ?$

STRATEGY: **Use the Commutative Property.**

STEP 1: Look at the numbers in the sum $287 + 462$. They are the reverse of the numbers $462 + 287$.

STEP 2: The Commutative Property states:

$$462 + 287 = 287 + 462$$

SOLUTION: **So, 287 + 462 = 749.**

THE COMMUTATIVE PROPERTY—MULTIPLICATION

The **Commutative Property** works for multiplication, too. You can multiply numbers in any order—the answer will be the same.

Here are two ways to show the Commutative Property in a multiplication problem:

1. $6 \times 7 = 42$
$7 \times 6 = 42$

2. $6 \times 7 = 7 \times 6 = 42$

EXAMPLE 2

$56 \times 28 = 1{,}568$. What is the answer to 28×56?

STRATEGY: **The Commutative Property for Multiplication states:**

$56 \times 28 = 28 \times 56$

SOLUTION: **So the answer for 28 × 56 is the same as the answer for 56 × 28. It is 1,568.**

NOTE: The Commutative Property does NOT work for subtraction or division! 4 – 3 is NOT equal to 3 – 4! And 4 ÷ 2 is NOT equal to 2 ÷ 4!

THE PROPERTY OF ZERO

There are two parts to this rule. One is for adding and one is for subtracting.

> The **Property of Zero for Adding**: If you add 0 to any number, then the sum is the number you started with.

Here are two ways to show the Property of Zero for Adding:

1. $64 + \mathbf{0} = 64$

2. $\mathbf{0} + 64 = 64$

> The **Property of Zero for Subtracting**: If you subtract 0 from any number, the answer is the number you started with.

Here is one way to show the Property of Zero for Subtracting:

$49 - \mathbf{0} = 49$

EXAMPLE 3

What number goes in the box in each of the following?

1. $53 + \square = 53$

2. $72 - 0 = \square$

3. $0 + 492 = \square$

4. $279 - \square = 279$

STRATEGIES: **Use the two parts (adding and subtracting) of the Property of Zero.**

1. Use the Property of Zero for Adding: $53 + \boxed{\mathbf{0}} = 53$

2. Use the Property of Zero for Subtracting: $72 - \mathbf{0} = \boxed{72}$

3. Use the Property of Zero for Adding: $\mathbf{0} + 492 = \boxed{492}$

4. Use the Property of Zero for Subtracting: $279 - \boxed{\mathbf{0}} = 279$

SAMPLE TEST QUESTIONS

1 95 + 0 = ☐

A 0 **B** 94

C 95 **D** 96

2 451 – 0 = ☐

A 0 **B** 400

C 450 **D** 451

3 0 + 991 = ☐

A 991 **B** 990

C 900 **D** 0

4 If 243 + 452 = 695, then 452 + 243 = ☐

A 695 **B** 452

C 243 **D** 209

5 0 + 0 = ☐

A 100 **B** 10

C 1 **D** 0

6 Which of the following is NOT true?

A 195 – 0 = 195

B 561 + 0 = 561

C 419 – 419 = 0

D 0 + 328 = 0

7 Which of the following is NOT true?

A 492 + 122 = 122 + 492

B 389 + 0 = 0

C 3,812 – 0 = 3,812

D 400 + 322 = 422 + 300

8 Since 67 × 84 = 5,628, then 84 × 67 = ☐

A 67 **B** 84

C 5,628 **D** 8,265

9 0 × 0 = ☐

A 1 **B** 0

C 10 **D** 2

EVEN AND ODD NUMBERS

The **even numbers** are: 0, 2, 4, 6, 8, 10, 12, and so forth.

The **odd numbers** are: 1, 3, 5, 7, 9, 11, 13, and so forth.

RULES FOR FINDING EVEN AND ODD NUMBERS

Even numbers end in: **0**, **2**, **4**, **6**, or **8**.

Odd numbers end in: **1**, **3**, **5**, **7**, or **9**.

EXAMPLE 1

Which of these is an odd number?

42, 58, 63, 70

STRATEGY: **Use the rules above.**

Look at the end of the numbers.

42, 58, and 70 are even numbers because they end in 2, 8, and 0.

SOLUTION: **63 is an odd number because it ends in 3.**

EXAMPLE 2

Is the number of blocks in this figure even or odd?

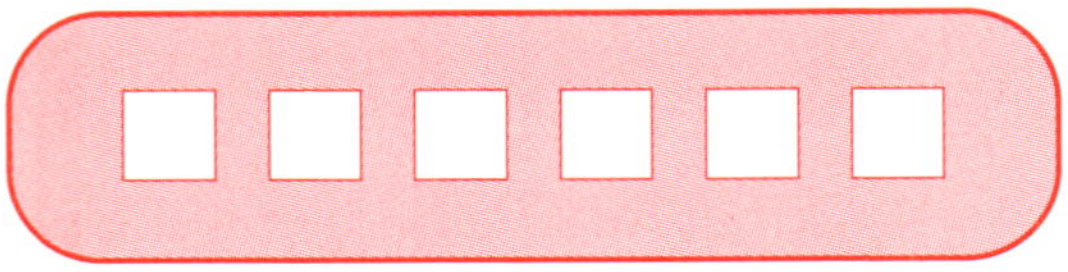

STRATEGY: **Count the blocks.**

There are 6 blocks.

SOLUTION: **Since 6 is an even number, the number of blocks is also an even number.**

EXAMPLE 3

How many odd numbers are there between 38 and 48?

STRATEGY: **Write all the numbers:**

STEP 1: Write the numbers from 38 to 48:

38, 39, 40, 41, 42, 43, 44, 45, 46, 47, 48

STEP 2: Underline the odd numbers, then count these odd numbers.

38, <u>**39**</u>, 40, <u>**41**</u>, 42, <u>**43**</u>, 44, <u>**45**</u>, 46, <u>**47**</u>, 48

SOLUTION: **There are 5 odd numbers between 38 and 48.**

SAMPLE TEST QUESTIONS

1 Which is an odd number?

A 62 **B** 70

C 75 **D** 96

2 Which is an even number?

A 17 **B** 28

C 35 **D** 49

3 Which of the following best describes this group of numbers: 6, 12, 14, 22, 38?

A They can all be divided by 4 evenly.

B They are all even.

C They are all odd.

D Some are even and some are odd.

4 Which of the following best describes this group of numbers: 5, 15, 17, 25, 29?

A They are all multiples of five.

B They are all even.

C They are all odd.

D Some are odd and some are even.

5 How many odd numbers are between 20 and 30?

A 3

B 4

C 5

D 6

6 Which of the following shows an odd number of blocks?

A 

B 

C 

D 

7 23,753 is an odd number because it ends in 3. Which of the following is true of the next counting number?

A It is even.

B It is odd.

C It ends in 0.

D There is not enough information to say.

LESSON 23

MODELS FOR FRACTIONS AND MIXED NUMBERS

FRACTIONS

A fraction is a part of a whole.

EXAMPLE 1

How much of the whole figure is shaded?

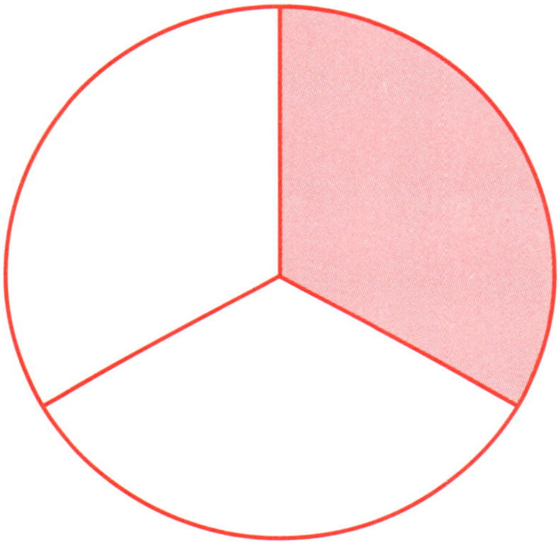

STRATEGY: **Follow these steps:**

STEP 1: Count the number of equal parts.

There are 3 equal parts.

STEP 2: Count the number of equal parts shaded.

There is 1 part shaded.

STEP 3: Form the fraction:

SOLUTION: $\frac{\textbf{Number of shaded parts}}{\textbf{Total number of parts}} = \frac{1}{3}$

EXAMPLE 2

What fraction names the shaded part of the set?

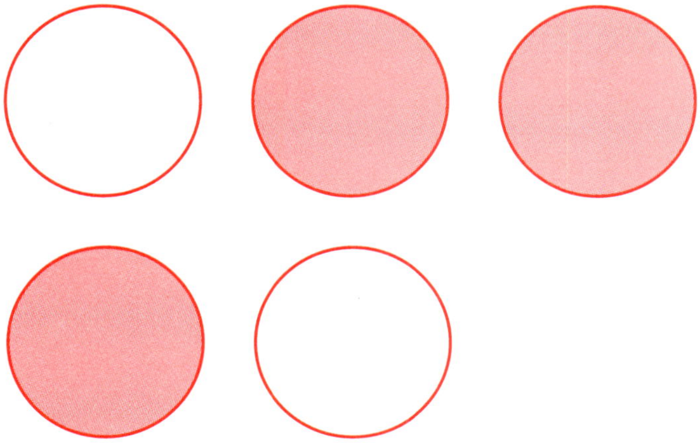

STRATEGY: **Count the circles.**

STEP 1: Count the total number of circles.

There are 5 circles.

STEP 2: Count the shaded circles.

There are 3 shaded circles.

STEP 3: Form the fraction:

SOLUTION: $\dfrac{\textbf{Number of shaded circles}}{\textbf{Total number of circles}} = \dfrac{3}{5}$

MIXED NUMBERS

A mixed number combines a whole number part and a fraction part.

EXAMPLE 3

This rectangle stands for a whole.

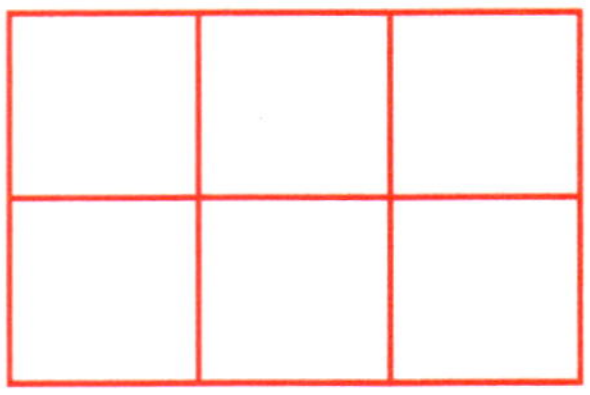

What part of the following is shaded?

Whole

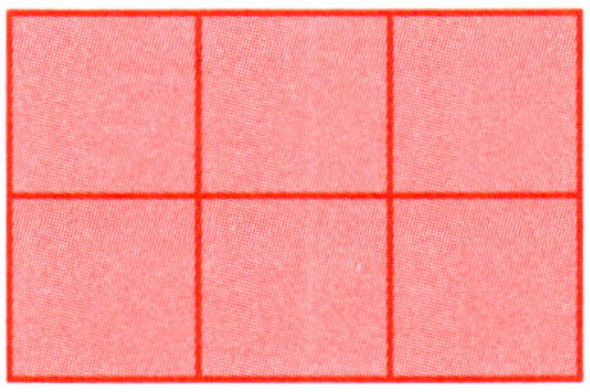

Whole

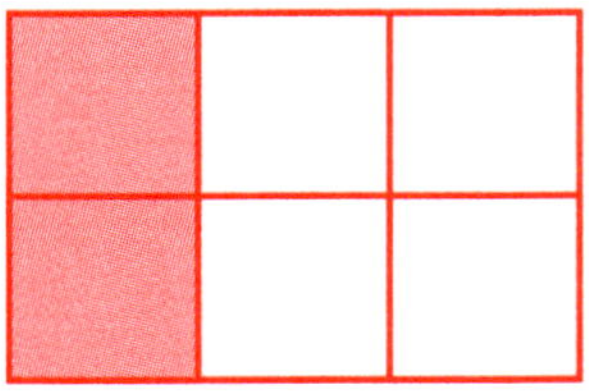

Fraction

STRATEGY: **Add the wholes and the fraction part.**

STEP 1: Count the whole parts.

There are 2 whole parts.

STEP 2: Name the fraction for the fraction part.

The fraction part is $\frac{2}{6}$.

STEP 3: Add the wholes and the fraction.

Number of wholes	2
Fraction	$+\frac{2}{6}$

SOLUTION: **The mixed number is $2\frac{2}{6}$.**
We say this number as "two and two-sixths."

SAMPLE TEST QUESTIONS

1 How much of the figure is shaded?

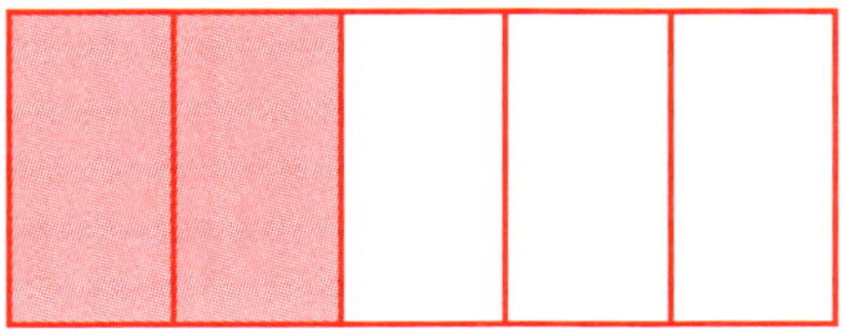

A $\frac{3}{5}$

B $\frac{2}{6}$

C $\frac{2}{5}$

D $\frac{3}{6}$

2 What fraction of the circles is shaded?

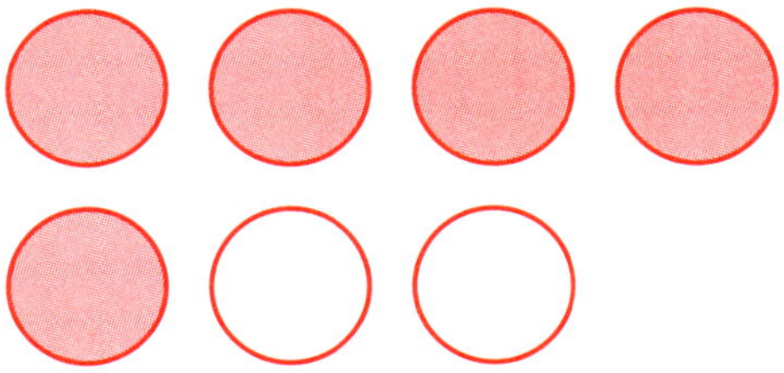

A $\frac{2}{7}$

B $\frac{5}{2}$

C $\frac{2}{5}$

D $\frac{5}{7}$

3 Which figure is $\frac{7}{10}$ shaded?

A

B

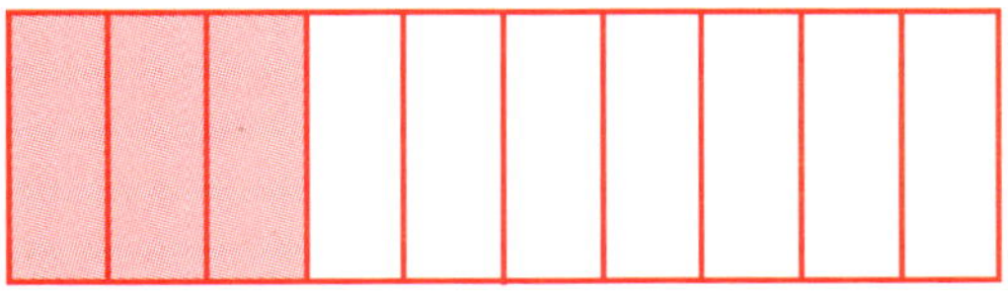

C

D

4 How much of this shape is shaded?

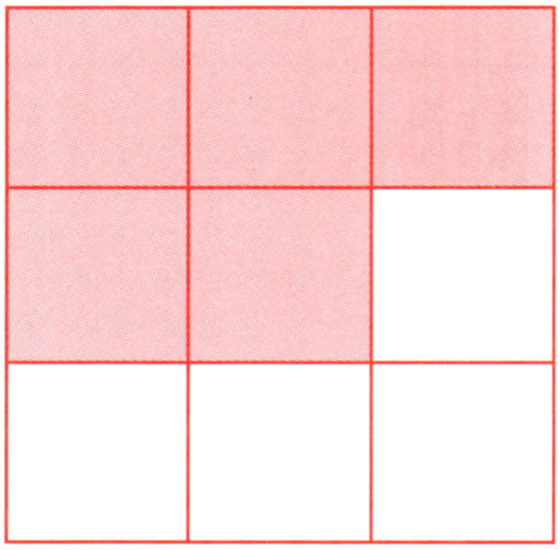

A $\frac{3}{9}$

B $\frac{5}{9}$

C $\frac{2}{3}$

D $\frac{5}{8}$

5 What fraction of the rectangle is shaded?

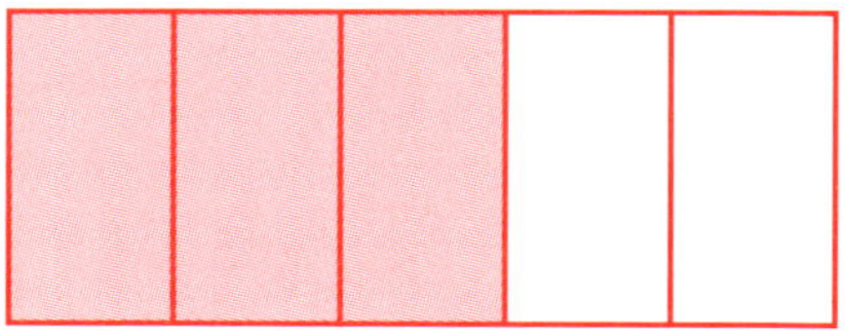

A $\frac{2}{5}$

B $\frac{3}{5}$

C $\frac{5}{3}$

D $\frac{2}{3}$

6 Choose the number for 3 wholes and $\frac{1}{3}$ of another whole.

A $\frac{1}{3}$

B 3

C $3\frac{1}{3}$

D $4\frac{1}{3}$

7 How do you say $5\frac{1}{6}$?

A Five and one-sixth

B Five-sixths and one-sixth

C Fifty-one-sixths

D Five and ten-sixths

8 The mixed number $2\frac{3}{4}$ means

A 2 wholes and 3 more

B 2 wholes and $\frac{3}{4}$ more

C $\frac{2}{4}$ and $\frac{3}{4}$

D 2 wholes and $\frac{1}{4}$ more

LESSON 24

FINDING THE FRACTIONAL PART OF A WHOLE

EXAMPLE 1

Patrick's school bought 24 computer disks. The computer teacher placed $\frac{1}{3}$ of the disks in Patrick's class. How many of these disks are in Patrick's class?

STRATEGY: **Use the meaning of $\frac{1}{3}$.**

STEP 1: What does $\frac{1}{3}$ mean?

The fraction $\frac{1}{3}$ means divide a set into 3 equal groups.

Then, each group is $\frac{1}{3}$.

STEP 2: Divide 24 disks into 3 equal groups.

Divide: $24 \div 3 = 8$

There are 8 disks in each of the 3 groups.

SOLUTION: **Patrick's class got 8 computer disks.**

EXAMPLE 2

Carla took a survey in her school. She found that $\frac{1}{7}$ of the students do not watch TV on school days. There are 238 students in her school. How many students do not watch TV on school days?

STRATEGY: **Use the meaning of $\frac{1}{7}$.**

STEP 1: What does $\frac{1}{7}$ mean?

The fraction $\frac{1}{7}$ means divide a set into 7 equal groups.

Then, each group is $\frac{1}{7}$.

STEP 2: Divide 238 by 7.

$$\begin{array}{r} 34 \\ 7\overline{)238} \\ -21 \\ \hline 28 \\ -28 \\ \hline 0 \end{array}$$

SOLUTION: **There are 34 students who do not watch TV on school days.**

SAMPLE TEST QUESTIONS

1 What is the correct way to find $\frac{1}{9}$ of 657?

A Multiply 657 by 9

B Divide 657 by 9

C Multiply 657 by 9 and then multiply by 2

D Divide 657 by 9 and then add 2

2 What is $\frac{1}{3}$ of 36?

A 10 **B** 12

C 16 **D** 32

3 Compute $\frac{1}{5}$ of 335.

A 67 **B** 68

C 69 **D** 75

4 Gloria is sending out 34 invitations for her birthday. So far, she has put stamps on $\frac{1}{2}$ of the envelopes. How many envelopes have stamps on them?

A 17 **B** 16

C 14 **D** 12

5 Billie estimates that he spends $\frac{1}{8}$ of each day eating. About how many hours does he eat each day? (Use 1 day = 24 hours.)

A 9 **B** 8

C 6 **D** 3

6 Aliza said that $\frac{1}{3}$ of the 600 pounds of paper can be recycled. How many pounds can be recycled?

A 100 **B** 200

C 300 **D** 333

7 David is a waiter in a restaurant where there are 45 tables. He is responsible for $\frac{1}{9}$ of the tables. How many tables is he responsible for?

A 9 **B** 6

C 5 **D** 3

8 Each period in Diego's school is 45 minutes long. When he looked at the clock during 4th period, he figured that the class was $\frac{2}{3}$ over. How many more minutes until the class was over?

A 9 **B** 11

C 15 **D** 30

LESSON 25

DECIMALS

The decimal **0.3** means the same as $\frac{3}{10}$.

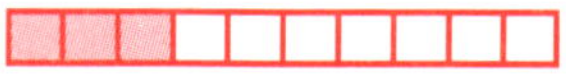

The decimal **0.23** means the same as $\frac{23}{100}$.

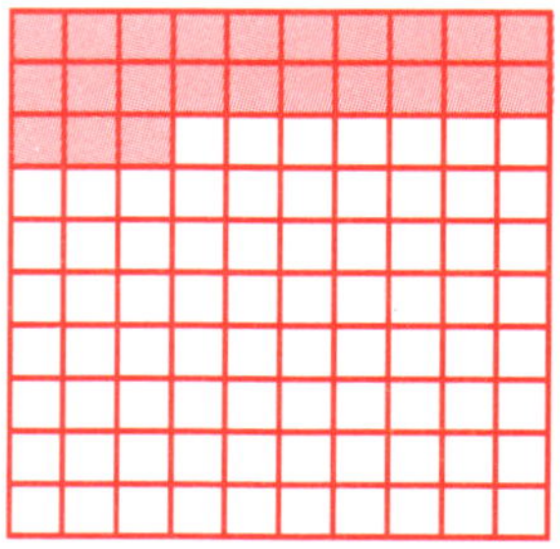

The decimal **2.45** means the same as $2\frac{45}{100}$.

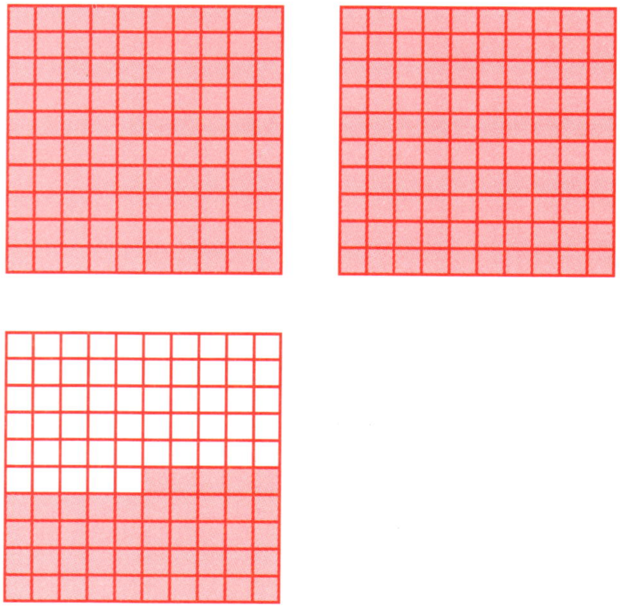

EXAMPLE 1

What is the decimal name for the shaded part of this figure?

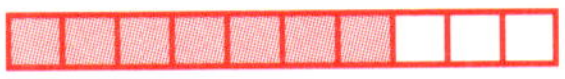

STRATEGY: **Write the fraction first.**

STEP 1: Write the fraction for the shaded part.

$$\frac{\text{Number of shaded parts}}{\text{Total number of parts}} = \mathbf{\frac{7}{10}}$$

STEP 2: Write the decimal for the fraction $\frac{7}{10}$.

SOLUTION: **The decimal for $\frac{7}{10}$ is 0.7**

EXAMPLE 2

Write the decimal represented by the shaded grids.

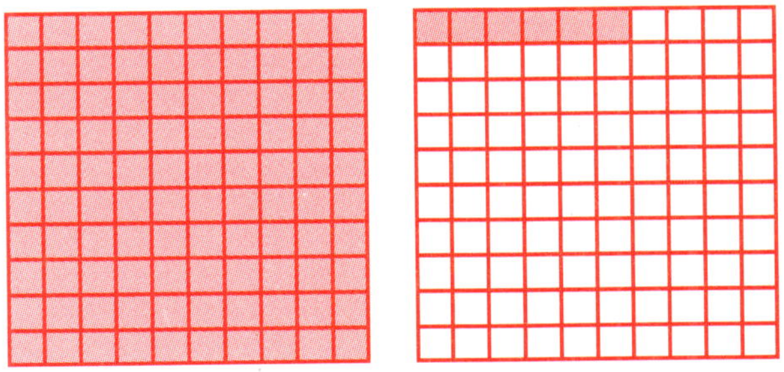

STRATEGY: **Write as a mixed number first.**

STEP 1: Write the mixed number for the shaded parts.

The shaded parts show 1 whole and $\frac{6}{100}$, or $1\frac{6}{100}$.

STEP 2: Write the decimal for $1\frac{6}{100}$.

SOLUTION: **The decimal for $1\frac{6}{100}$ is 1.06**

EXAMPLE 3

What does the **3** in 2.4**3** stand for?

STRATEGY: **Use a place value chart.**

STEP 1: Write 2.43 in a place value chart.

Ones	Tenths	Hundredths
2	4	3

STEP 2: Read the value of the 3 from the chart.

SOLUTION: **The 3 stands for 3 hundredths.**

SAMPLE TEST QUESTIONS

1 Choose the decimal represented by the shaded grids.

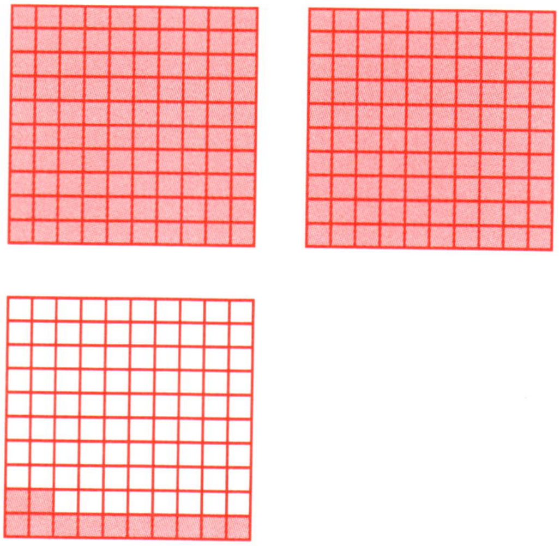

A 2.02 **B** 0.14

C 1.12 **D** 2.12

2 George lifted a box weighing 8.47 kilograms. What does the 7 stand for?

A 7

B 0.7

C 0.07

D 0.007

3 Which decimal is largest?

A 0.06 **B** 0.006

C 0.6 **D** 0.095

4 Which decimal is the same as $3\frac{7}{100}$?

A 3.7 **B** 3.07

C 0.37 **D** 0.037

5 How much more is 6.27 than 6.17?

A 10

B 1

C 0.1

D 0.01

6 What does the 4 stand for in 85.64?

A 4 hundreds

B 4 tens

C 4 tenths

D 4 hundredths

7 Which decimal is between 0.7 and 0.8?

A 0.65 **B** 0.75

C 0.85 **D** 0.95

8 What is the correct way to write the number represented below?

4 tenths
2 tens
7 hundredths
3 ones

A 2347.0

B 234.7

C 23.47

D 2.347

9 What decimal is 0.01 more than 6.99?

A 7.01

B 7.00

C 6.98

D 6.97

LESSON 26

FRACTION-DECIMAL EQUIVALENCES AND ORDERING FRACTIONS AND DECIMALS

You should know these **equivalences**. "Equivalences" means "things that are the same as each other."

Fraction	Decimal
$\frac{1}{100}$	0.01
$\frac{1}{10}$	0.1
$\frac{1}{4}$	0.25
$\frac{1}{2}$	0.5
$\frac{3}{4}$	0.75

EXAMPLE 1

What decimal is the same as $\frac{3}{12}$?

STRATEGY: **Simplify the fraction, then convert it to a decimal.**

STEP 1: Simplify the fraction.

Divide the numerator and denominator by 3.

$$\frac{3}{12} = \frac{3 \div 3}{12 \div 3} = \frac{1}{4}$$

STEP 2: Change $\frac{1}{4}$ to a decimal. (Use the chart at the top of the page.)

$$\frac{1}{4} = 0.25$$

SOLUTION: **The decimal that is equivalent to $\frac{3}{12}$ is 0.25**

EXAMPLE 2

Write these numbers in order from least to greatest: 2.05, 0.03, $\frac{1}{4}$, $2\frac{1}{2}$

STRATEGY: **Compare the numbers as decimals.**

STEP 1: Change the fraction and mixed number to decimals.

Use the chart at the beginning of this lesson.

$\frac{1}{4} = 0.25$

$2\frac{1}{2} = 2.5$

STEP 2: Put the numbers in a place value chart.

	ones	tenths	hundredths
2.05 =	2	0	5
0.03 =	0	0	3
$\frac{1}{4} = 0.25 =$	0	2	5
$2\frac{1}{2} = 2.5 =$	2	5	0

STEP 3: Use the values of the places to order the numbers from least to greatest.

SOLUTION: **The order of the numbers from least to greatest is: 0.03, $\frac{1}{4}$, 2.05, and $2\frac{1}{2}$**

EXAMPLE 3

Which is larger—4.65 or 4.56?

STRATEGY: **Use place value to compare these numbers.**

STEP 1: Put the numbers in a place value chart.

	ones	tenths	hundredths
4.65 =	4	6	5
4.56 =	4	5	6

STEP 2: Compare the values of the places.

4.65 and 4.56 both have 4 ones.

4.65 has 6 tenths and 4.56 has only 5 tenths.

SOLUTION: **4.65 is larger than 4.56**

SAMPLE TEST QUESTIONS

1 Which list shows these numbers from least to greatest?

4.05, 0.07, $\frac{1}{4}$, $4\frac{3}{4}$

A $\frac{1}{4}$, 0.07, $4\frac{3}{4}$, 4.05

B $\frac{1}{4}$, 0.07, 4.05, $4\frac{3}{4}$

C 0.07, $\frac{1}{4}$, $4\frac{3}{4}$, 4.05

D 0.07, $\frac{1}{4}$, 4.05, $4\frac{3}{4}$

2 Jill saw this sign on the highway.

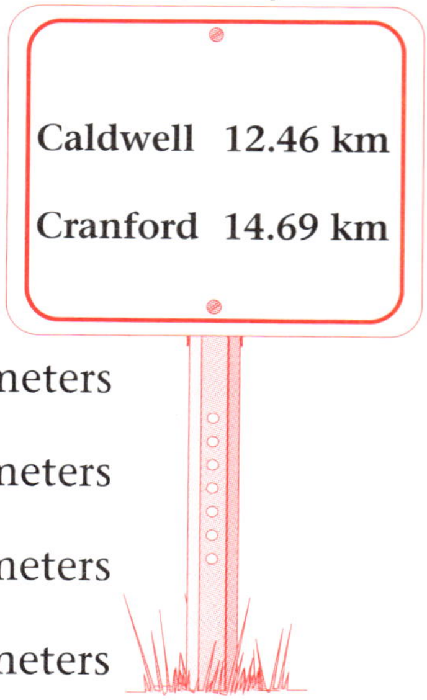

How much farther is Cranford than Caldwell?

A 2.23 kilometers

B 2.13 kilometers

C 1.23 kilometers

D 1.13 kilometers

3 What decimal stands for $\frac{6}{12}$?

A 0.25

B 0.5

C 0.1

D 0.75

4 One penny is equal to 0.01 of a dollar. Which of these decimals equals 35 pennies?

A 0.0035

B 0.035

C 0.305

D 0.35

5 What decimal is equivalent to $7\frac{9}{12}$?

A 7.75

B 7.9

C 7.95

D 7.99

6 Which number sentence is true?

A $0.07 > 0.17$

B $0.34 < 0.43$

C $0.2 > 0.23$

D $0.44 < 0.4$

7 Last week, Sean weighed 30.23 kilograms. He gained $\frac{1}{2}$ kilogram. How much does he weigh now?

A 30.28 kilograms

B 30.73 kilograms

C 31.23 kilograms

D 35.23 kilograms

8 Which set of numbers is in order from least to greatest?

A $3.8, 3.81, 3\frac{9}{10}, 3\frac{9}{100}$

B $3\frac{9}{10}, 3.8, 3.81, 3\frac{9}{100}$

C $3\frac{9}{100}, 3.8, 3\frac{9}{10}, 3.81$

D $3\frac{9}{100}, 3.8, 3.81, 3\frac{9}{10}$

LESSON 27 ROUNDING TO THE NEAREST 10, 100, AND 1,000

ROUNDING TO THE NEAREST 10

EXAMPLE 1

Round 67 to the nearest 10.

STRATEGY: **Follow these steps.**

STEP 1: Place 67 on a number line marked off in 10's.

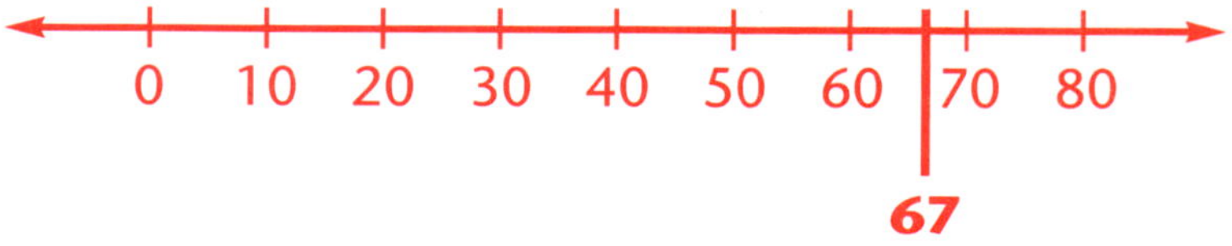

STEP 2: Decide if 67 is closer to 60 or 70.

67 is closer to 70. So round UP.

SOLUTION: **Since 67 is closer to 70, round 67 up to 70.**

EXAMPLE 2

Round 35 to the nearest 10.

STRATEGY: **Use the number line.**

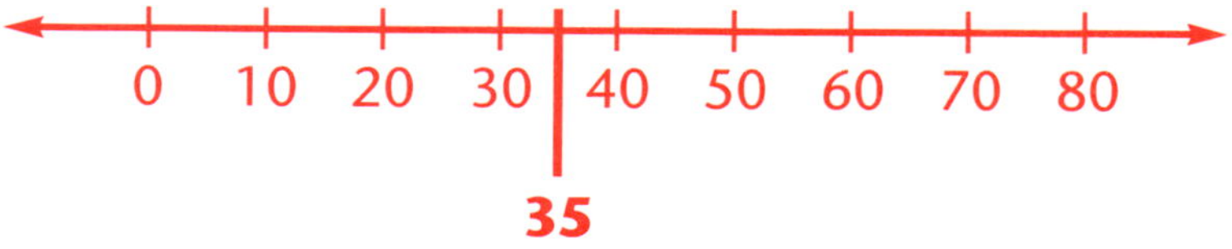

35 is halfway between 30 and 40. When this happens, round UP.

SOLUTION: **Round 35 up to 40.**

ROUNDING TO THE NEAREST 100

EXAMPLE 3

Round 417 to the nearest 100.

STRATEGY: **Use the number line.**

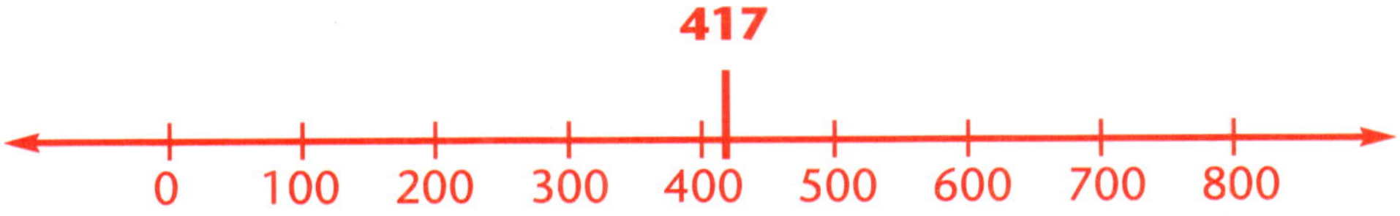

Decide if 417 is closer to 400 or 500.

SOLUTION: **Since 417 is closer to 400 than 500, round DOWN to 400.**

EXAMPLE 4

Round 350 to the nearest 100.

STRATEGY: **Use the number line as a guide.**

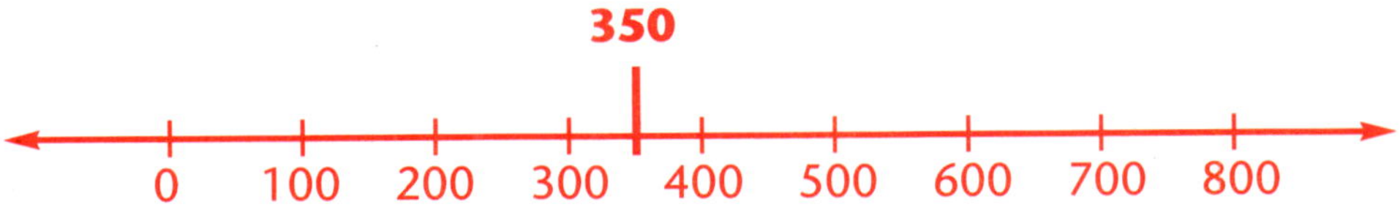

350 is halfway between 300 and 400.

When a number ends in 50, round UP.

SOLUTION: **Round 350 up to 400.**

ROUNDING TO THE NEAREST 1,000

EXAMPLE 5

Round 5,615 to the nearest 1,000.

STRATEGY: **Use the number line strategy.**

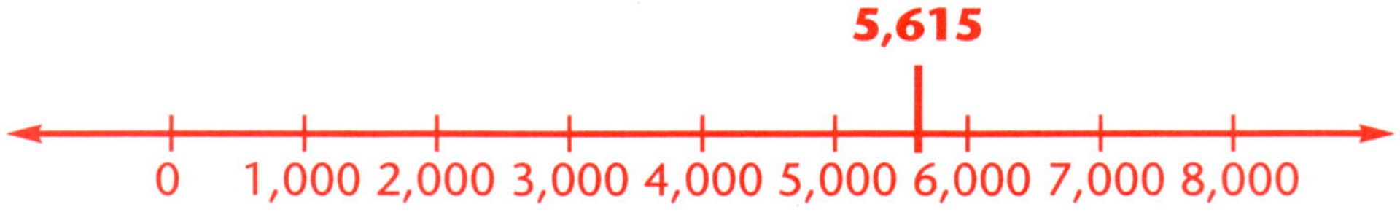

5,615 is closer to 6,000 than to 5,000. So round UP.

SOLUTION: **Round 5,615 up to 6,000.**

EXAMPLE 6

Round 5,500 to the nearest 1,000.

STRATEGY: **Use the number line strategy.**

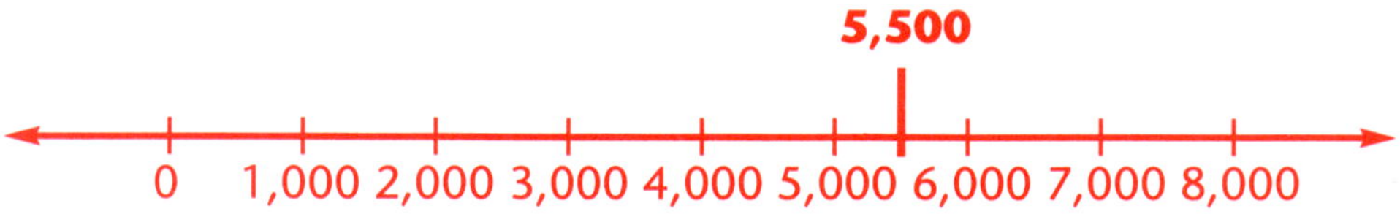

5,500 is halfway between 5,000 and 6,000.

When a number ends in 500, round UP.

SOLUTION: **Round 5,500 up to 6,000.**

SAMPLE TEST QUESTIONS

1 A new TV set is advertised as selling for $424. Round this figure to the nearest 10 dollars.

A $400

B $410

C $420

D $430

2 The distance from John's house to his grandparents' home is 114 miles. About how many miles is this distance? (Round this number to the nearest 100 miles.)

A 200 miles

B 150 miles

C 100 miles

D 50 miles

3 Which number does not round to 500 when rounded to the nearest hundred?

A 451

B 490

C 524

D 550

4 Stan's bill at the bicycle shop came to $73.24. The owner rounded the amount to the nearest $10. What is true about the rounded amount?

A It is less than $73.24.

B It is more than $73.24.

C It is exactly equal to $73.24.

D It is $10 less than $73.24.

5 Mark counted 4,803 tickets sold for the Fall Festival. Round this figure to the nearest hundred.

A 5,000

B 4,900

C 4,800

D 4,000

6 Round 2,500 to the nearest 1,000.

A 2,000

B 3,000

C 3,500

D 5,000

LESSON 28

USING ESTIMATION TO SOLVE PROBLEMS

An **estimate** is an answer that is not exact. It is close to an exact answer.

We make estimates for distances, amounts of money, weights, and for many other things.

EXAMPLE 1

The closest estimate of 8,151 + 3,902 is _______

A 5,000

B 10,000

C 11,000

D 12,000

STRATEGY: **Round each number and add.**

STEP 1: Round 8,151 to the nearest 1,000.

8,151 rounded to the nearest 1,000 is 8,000.

STEP 2: Round 3,902 to the nearest 1,000.

3,902 rounded to the nearest 1,000 is 4,000.

STEP 3: Add the rounded numbers:

8,000 + 4,000 = 12,000

SOLUTION: **The answer that is the closest estimate is Answer D—12,000.**

EXAMPLE 2

Nicki earns $410 each week. About how much does she earn in a year?

STRATEGY: Use rounded numbers to make estimates.

STEP 1: Round $410 to the nearest 100 dollars.

$410 rounded to the nearest hundred dollars is $400.

STEP 2: How many weeks are there in a year?

There are 52 weeks in a year.

STEP 3: Round the number of weeks in a year to the nearest 10.

52 rounded to the nearest 10 is 50.

STEP 4: Multiply the rounded numbers:

$$\begin{array}{r} 400 \\ \times \quad 50 \\ \hline 20{,}000 \end{array}$$

SOLUTION: Nicki earns about $20,000 in a year. This is an estimate.

SAMPLE TEST QUESTIONS

1 The closest estimate of $456 – $203 is ______

A $200 **B** $230

C $260 **D** $300

2 Tom saves $45 each month. What is the best estimate of how much he will he save at the end of one year?

A $300 **B** $400

C $500 **D** $700

3 About how many minutes are there in 1 day (24 hours)?

A 150

B 1,500

C 15,000

D 150,000

4 About how many hours are there in 1 year?

A 90 **B** 900

C 9,000 **D** 90,000

5 A new machine can print postage on envelopes at the rate of 725 envelopes per hour. What is the best estimate of how many envelopes the machine can do in 8 hours?

A 6,000 **B** 5,000

C 4,000 **D** 3,000

6 The closest estimate of 14,884 + 12,932 is ______ .

A 2,000 **B** 26,000

C 27,000 **D** 28,000

7 Vanessa's class uses the computer lab in her school for 28 minutes each day. What is a good estimate of how long her class uses the computer lab in 22 days (about the number of school days per month)?

A 8 hours

B 10 hours

C 12 hours

D 14 hours

LESSON 29

ROUNDING DECIMALS AND FRACTIONS

EXAMPLE 1

Juan said his parents' living room is 4.7 meters long. Round 4.7 meters to the nearest whole number.

STRATEGY: **Find the whole number that is closest to 4.7.**

STEP 1: Place 4.7 on a number line marked off in intervals of 0.5:

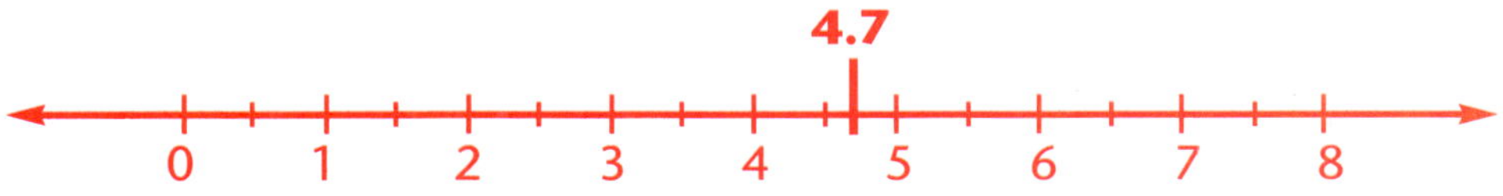

STEP 2: Ask: is 4.7 closer to 4 or 5?

4.7 is closer to 5.

SOLUTION: **4.7 rounded to the nearest whole number is 5. The living room is about 5 meters long.**

EXAMPLE 2

Paul took 17.5 hours to finish a project at work. Round 17.5 to the nearest whole number.

STRATEGY: **Use the following rule: If a number is exactly halfway between two whole numbers, round UP.**

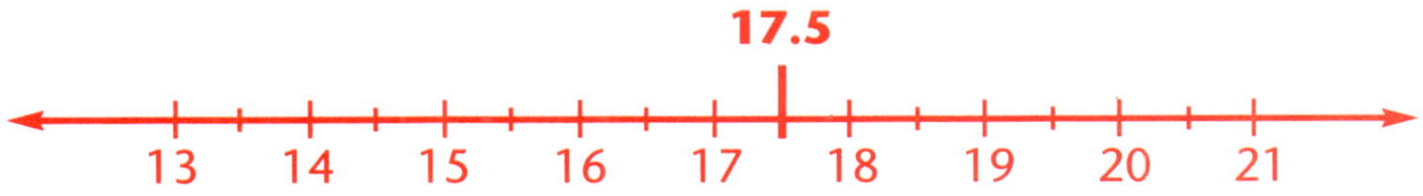

17.5 is exactly halfway between 17 and 18. In this case, round UP to 18.

SOLUTION: **17.5 rounded to the nearest whole number is 18. So Paul took about 18 hours to finish the project.**

EXAMPLE 3

Donna has $6\frac{1}{4}$ feet of ribbon. Round $6\frac{1}{4}$ to the nearest whole number.

STRATEGY: **Find the whole number that $6\frac{1}{4}$ is closest to.**

STEP 1: Between which two whole numbers is $6\frac{1}{4}$?

$6\frac{1}{4}$ is between 6 and 7.

STEP 2: Is $6\frac{1}{4}$ closer to 6 or to 7?

Since $\frac{1}{4} < \frac{1}{2}$, then $6\frac{1}{4}$ is closer to 6.

SOLUTION: **$6\frac{1}{4}$ rounded to the nearest whole number is 6. Donna has about 6 feet of ribbon.**

EXAMPLE 4

Andy used $23\frac{1}{2}$ cups of flour to make a cake for a banquet. Round 23 to the nearest whole number.

STRATEGY: **Use the rule: When a mixed number ends in $\frac{1}{2}$ (halfway between two whole numbers), round UP.**

$23\frac{1}{2}$ ends in $\frac{1}{2}$. So round UP.

SOLUTION: **$23\frac{1}{2}$ rounded to the nearest whole number is 24. Andy used about 24 cups of flour.**

SAMPLE TEST QUESTIONS

1 Julie traveled 65.5 miles to get to the stadium. Round 65.5 to the nearest whole number.

A 60

B 65

C 66

D 70

2 Phil watched TV for $6\frac{1}{4}$ hours over the weekend. Round this time to the nearest whole number.

A 6 hours

B $6\frac{1}{2}$ hours

C 7 hours

D 10 hours

3 Which of these fractions is closest to 4?

A $3\frac{1}{2}$

B $3\frac{3}{4}$

C $4\frac{4}{5}$

D $4\frac{1}{2}$

4 Michael's new fishing pole is $7\frac{2}{8}$ feet long. Round to the nearest whole number.

A $\frac{29}{4}$ feet

B 7 feet

C $7\frac{1}{4}$ feet

D 10 feet

5 Which of these fractions is less than $\frac{1}{2}$?

A $\frac{2}{4}$

B $\frac{6}{9}$

C $\frac{1}{3}$

D $\frac{9}{10}$

6 Norma measured the length of her table. It is 1.7 meters long. Round this length to the nearest whole number.

A 1 meter

B 1.5 meters

C 1.8 meters

D 2 meters

REVIEW OF LESSONS 20-29

1 Which of the following is NOT true?

A 455 + 0 = 455

B 455 – 0 = 0

C 455 – 455 = 0

D 0 + 0 = 0

2 □ + 0 = 0

A 0

B 1

C 10

D 100

3 How many odd numbers are between 24 and 34?

A 2

B 3

C 4

D 5

4 What is the value of "8" in 38,621?

A 800,000

B 80,000

C 8,000

D 800

5 If you are rounding to the nearest 1,000, which number does not round to 2,000?

A 1,499

B 1,500

C 1,898

D 2,499

6 What fraction is shaded?

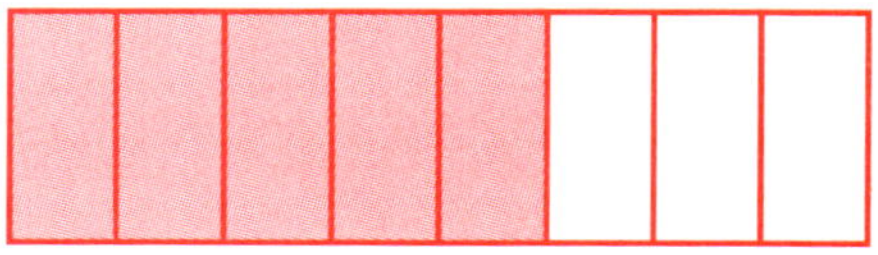

A $\frac{3}{8}$ **B** $\frac{5}{8}$

C $\frac{4}{8}$ **D** $\frac{6}{8}$

7 Find $\frac{1}{8}$ of 48.

A 4

B 6

C 9

D 16

8 Choose the decimal represented by the shaded grid.

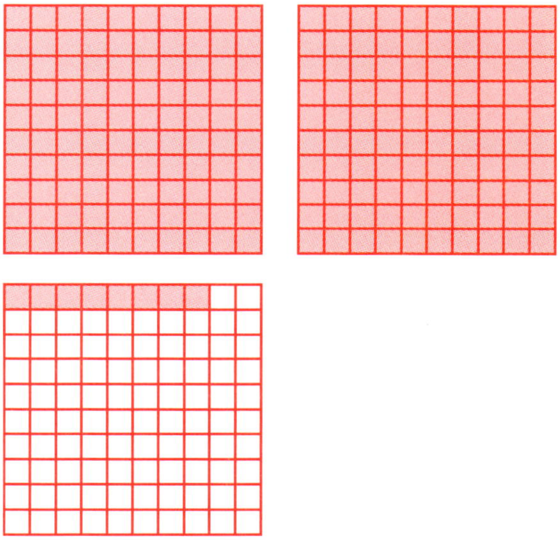

A 0.028

B 0.28

C 2.08

D 2.8

9 Which number sentence is true?

A $0.07 > 0.7$

B $0.85 < 0.58$

C $0.6 < 0.64$

D $0.88 < 0.8$

10 The closest estimate of 78×22 is ________ .

A 160

B 1,600

C 16,000

D 160,000

11 Which of these numbers is the closest to 2?

A 1.75

B $2\frac{1}{4}$

C 2.05

D $1\frac{8}{10}$

OPEN-ENDED QUESTIONS

1 I am a mystery number. Use the following clues to find all the numbers that I could be. Show how you got your answer.

Clues

- I am even.
- I am a 2-digit number.
- The sum of my digits is 8.

2 a) Choose a number greater than 10,000 and less than 999,999 that has at least five different digits: ____________

b) Write the number in expanded form (for example, 12,418 would be 10,000 + 2,000 + 400 + 10 + 8): ________________________

c) Using the symbols below, write that same number:

____

a = 100,000	c = 1,000	e = 10
b = 10,000	d = 100	f = 1

3 Draw a rectangle and divide it into 8 equal parts. Use this figure to explain why $\frac{1}{4} = \frac{2}{8}$.

ADDING AND SUBTRACTING FRACTIONS WITH LIKE DENOMINATORS

Two fractions have **like denominators** when their denominators are the same. The fractions $\frac{4}{7}$ and $\frac{1}{7}$ have like denominators, since both denominators are 7.

Fractions with like denominators are the easiest fractions to add. **All you have to do is add numerators**.

And when you **subtract** fractions with like denominators, just **subtract the numerators**.

EXAMPLE 1

Marty and Hans have the job of mowing the football field at school. Marty mowed $\frac{3}{10}$ of the field. Hans mowed $\frac{1}{10}$ of the field. What fraction of the field did they mow together?

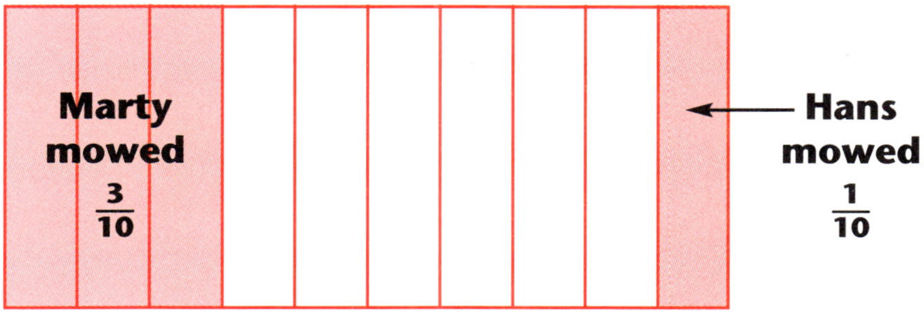

STRATEGY: **Check to see if the fractions have like denominators. If they do, add the numerators.**

STEP 1: What are the denominators of the two fractions?

Both fractions have the denominator 10.

STEP 2: Add the numerators:

$$\frac{3}{10} + \frac{1}{10} = \frac{4}{10}$$

SOLUTION: **Marty and Hans mowed $\frac{4}{10}$ of the football field.**

SAMPLE TEST QUESTIONS

1 Joanna ate $\frac{1}{8}$ of a pizza and her sister Dana ate $\frac{3}{8}$ of the same pizza. What fraction of the pizza did they eat?

A $\frac{2}{8}$

B $\frac{4}{8}$

C $\frac{1}{8}$

D $\frac{3}{8}$

2 $\frac{3}{13} + \frac{2}{13} = ?$

A $\frac{5}{26}$

B $\frac{1}{13}$

C $\frac{4}{13}$

D $\frac{5}{13}$

3 Waldo took $\frac{1}{6}$ of the stickers from the pack and Alicia took $\frac{3}{6}$ of the stickers. How much more (as a fraction) of the pack did Alicia take?

A $\frac{4}{6}$

B $\frac{2}{3}$

C $\frac{2}{6}$

D $\frac{1}{6}$

4 $\frac{7}{13} - \frac{1}{13} = ?$

A $\frac{6}{26}$

B $\frac{1}{13}$

C $\frac{4}{13}$

D $\frac{6}{13}$

5 Georgia spent $\frac{1}{3}$ of an hour on a science report and another $\frac{1}{3}$ of an hour on a social studies report. What fraction of an hour did she spend on these reports?

A $\frac{2}{6}$

B $\frac{2}{5}$

C $\frac{2}{3}$

D $\frac{4}{3}$

6 $\frac{5}{9} + \frac{2}{9} = ?$

A $\frac{3}{9}$

B $\frac{4}{9}$

C $\frac{6}{9}$

D $\frac{7}{9}$

7 Three-fifths of the students in the third grade walk to school. One-fifth of the students travel by bus. What fraction of the students either walk or come to school by bus?

A $\frac{4}{10}$

B $\frac{2}{5}$

C $\frac{3}{5}$

D $\frac{4}{5}$

8 $\frac{9}{11} - \frac{1}{11} = ?$

A $\frac{10}{11}$

B $\frac{9}{11}$

C $\frac{8}{11}$

D $\frac{7}{11}$

9 Doug has $\frac{4}{9}$ of the answers on his history test correct. Fran has $\frac{7}{9}$ of the answers correct. How much more (as a fraction) does Fran have correct?

A $\frac{3}{9}$

B $\frac{2}{9}$

C $\frac{11}{18}$

D $\frac{11}{9}$

10 $\frac{4}{13} - \frac{4}{13} = ?$

A $\frac{8}{13}$

B $\frac{2}{13}$

C $\frac{1}{13}$

D 0

LESSON 31

ADDING AND SUBTRACTING DECIMALS WITH TENTHS

The decimal 0.6 means the same as $\frac{6}{10}$.

This diagram shows 0.6 shaded.

It means 6 of 10 equal parts.

Read 0.6 the same way that you read.

You say "6 tenths."

HOW TO ADD TWO DECIMALS

1. Write the numbers under each other.
2. Make sure the decimal points are lined up.
3. Add each place, starting at the right.
4. Place the decimal point in the sum, lined up under the other decimal points.

EXAMPLE

Gil drove 5.3 miles on Monday and 6.1 miles on Tuesday. How far did he drive all together on these two days?

STRATEGY: **Follow the steps above.**

$$\begin{array}{r} 5.3 \\ +\ 6.1 \\ \hline 11.4 \end{array}$$

SOLUTION: **Gil drove 11.4 miles in the two days.**

SAMPLE TEST QUESTIONS

1 3.6 + 2.9 = ?

A 5.5

B 5.8

C 6.4

D 6.5

2 Marian spent 2.4 hours writing a report and 0.8 hour editing the report. How much time did it take her to complete the report?

A 3.1 hours

B 3.2 hours

C 3.8 hours

D 4.2 hours

3 Dwight drove 9.3 miles to visit his brother, then drove 2.5 miles to see his sister. How many miles did he drive all together?

A 11.5

B 11.8

C 12.3

D 12.5

4 Melba slept 7.3 hours on Tuesday and 6.2 hours on Wednesday. How many hours did she sleep on these two days?

A 13.5

B 13.4

C 13.2

D 12.5

5 Susie walked 10.4 km and ran 4.6 km yesterday. How many kilometers did she cover?

A 14

B 14.5

C 14.8

D 15

6 Mario has two puppies. One weighs 3.4 pounds and the other weighs 4.7 pounds. How much do the two puppies weigh together?

A 8.3 pounds

B 8.2 pounds

C 8.1 pounds

D 7.1 pounds

LESSON 32

COMPLETING NUMBER SENTENCES

If you know addition facts, you know the answer to 6 + 3 = □

But do you know the missing number in 5 + □ = 12?

EXAMPLE 1

Find the missing number in 5 + □ = 12.

STRATEGY: **Use addition facts.**

Think: what number added to 5 equals 12?

SOLUTION: **Since 5 + 7 = 12, then the missing number is 7.**

EXAMPLE 2

What is the missing number?

□ × 4 = 12

STRATEGY: **Make sure you see this as multiplication.**

STEP 1: What does □ × 4 = 12 mean?

It means: how many 4's are there in 12?

STEP 2: Answer this question:

Since 4 + 4 + 4 = 12, then three 4's equal 12.

3 × 4 = 12

SOLUTION: **The missing number is 3.**

EXAMPLE 3

Choose the missing sign.

6 ☐ 4 = 24

A $+$

B $-$

C $\times$

D $\div$

STRATEGY: **Try each sign.**

If you try $+$, $-$, $\times$, and $\div$, you will find that only $\times$ works.

$6 \times 4 = 24$

SOLUTION: **So, $\times$ is the missing sign. The answer is C.**

SAMPLE TEST QUESTIONS

Find the missing number for Questions 1–6.

1 $6 + \square = 10$

A 3

B 4

C 5

D 6

2 $11 - \square = 8$

A 6

B 5

C 4

D 3

3 $2 \times \square = 8$

A 4

B 3

C 2

D 1

4 $4 + \square = 13$

A 9

B 8

C 7

D 6

5 $7 \times \square = 28$

A 3

B 4

C 5

D 6

6 $12 - \square = 8$

A 4

B 5

C 6

D 7

Find the missing sign for Questions 7–10.

7 $5 \square 7 = 12$

A $+$

B $-$

C $\times$

D $\div$

8 $6 \square 5 = 30$

A $+$

B $-$

C $\times$

D $\div$

9 $2 + 4 + 1 + 9 = 20 \square 4$

A $-$

B $=$

C $\times$

D $\div$

10 $10 \times 2 \square 4 \times 5$

A $+$

B $=$

C $-$

D $\div$

LESSON 33 SOLVING PROBLEMS WITH NUMBER SENTENCES

You should know how a number sentence can be used to solve a problem.

EXAMPLE 1

Connie spent $8 in the morning and $7 in the afternoon. Which number sentence shows how much she spent all together?

A $8 + 7 = \square$

B $8 - 7 = \square$

C $8 \times 7 = \square$

D $8 + 8 = \square$

STRATEGY: **Use key words in the problem.**

STEP 1: Find the key words in the question.

The key words are **all together**.

STEP 2: What is the meaning of **all together**?

All together means that you should add.

SOLUTION: **The answer is A.**

EXAMPLE 2

There are 4 teams in the basketball tournament. Each team has 8 students on it. Which number sentence shows how many students are on the 4 teams?

A $8 \times 8 = \square$

B $8 + 4 = \square$

C $4 \times 8 = \square$

D $4 \times 4 = \square$

STRATEGY: **Think of what operation to use.**

STEP 1: What operation do you use—addition, subtraction, multiplication, or division?

Think: 4 teams of 8 students.

The fast way to get the answer is to multiply.

STEP 2: What do you multiply?

You multiply 4×8, which is the same as adding 8 four times.

(You can solve the problem by adding: $8 + 8 + 8 + 8$, but multiplying is quicker.)

SOLUTION: **The solution is C.**

SAMPLE TEST QUESTIONS

1 Sue bought 12 donuts. Each donut cost 45 cents. Which number sentence can you use to find the total cost in cents?

A $12 \times 45 =$

B $12 + 45 =$

C $45 - 12 =$

D $45 + 45 + 45 =$

2 Choose the story problem that goes with the number sentence:

$6 + 3 = 9$

A Nancy called 6 people on Tuesday, 3 people on Wednesday, and 9 more on Thursday. How many people did she call on those 3 days?

B Al called 6 people a day for 3 days. How many people did he call?

C Carlos called 6 people on Tuesday, 2 people on Wednesday, and 3 people on Thursday. How many people did he call on those 3 days?

D Todd called 6 people on Tuesday and 3 more on Wednesday. How many people did he call?

3 At the Fun Ride Festival, you have to pay $4 for the first ride, then you pay $2 for each ride after that. Which of the following could be used to find the cost of 4 rides?

A $4 + 3 + 2$

B $3 + 2 + 2 + 3$

C $4 + 2 + 2$

D $4 + 2 + 2 + 2$

4 Choose the story problem that goes with the number sentence: **$18 - 12 = 6$**

A Nora bought 18 tickets in the morning and 12 tickets in the afternoon. How many tickets did she buy?

B Jeff bought 18 tickets each morning for 12 days. How many tickets did he buy?

C Selene bought 18 tickets in the morning and gave 12 tickets to her classmates. How many tickets did she have left?

D Julie bought 18 tickets. She divided the tickets among 6 students. How many tickets did each get?

LESSON 34

STRATEGIES FOR PROBLEM SOLVING

You should know about these strategies for solving problems.

STRATEGIES

1. Act it out
2. Make a model
3. Draw a picture
4. Make a chart or graph
5. Look for a pattern
6. Make a simpler problem
7. Use logic
8. Work backwards
9. Guess and check
10. Break into parts

EXAMPLE

Carlos lives 6 miles from Henry, and Henry lives 4 miles from Georgia. Which of the following is NOT a possible distance between Carlos and Georgia?

A 2 miles

B 4 miles

C 10 miles

D 12 miles

STRATEGY: Choose the strategy that best fits this problem.

STEP 1: Read the problem several times. Then look at the list of strategies on the previous page. Ask yourself: Which strategy would help you solve this problem?

One of the best strategies for this type of problem is "**Draw a picture**." This is a good strategy since you have a better chance of solving the problem if you see the three locations on paper.

You will have to draw several pictures since there is more than one way that the houses can be set up.

STEP 2: Draw a picture of the three places.

Picture 1

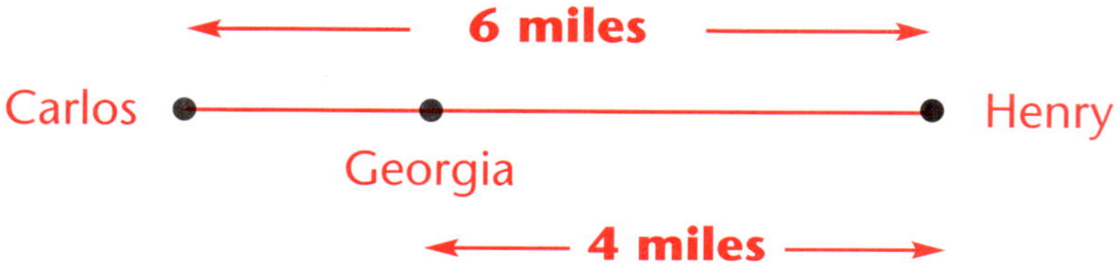

Picture 1 shows us that the 3 places can be on the same line. Here, Georgia lives 2 miles from Carlos ($6 - 4 = 2$).

Answer A is possible.

STEP 3: Draw another picture.

Picture 2

Picture 2 shows us another way that the locations can be on the same line. In this case, the distance between Georgia and Carlos is 10 miles. Answer C is possible.

This is also the farthest Carlos can be from Georgia.

STEP 4: Draw another picture.

Picture 3

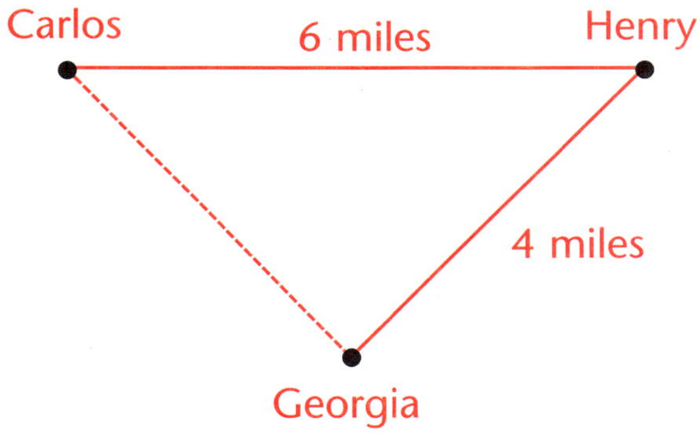

From Picture 3, you can see that the distance from Carlos to Georgia could be 4 miles. So Answer B is possible.

We eliminated Answers A, B, and C—all are possible.

STEP 5: Ask: Do you need to draw any more pictures?

No. We've already eliminated answers A, B, and C. We can't eliminate answer D, since 10 miles was the farthest Carlos could be from Georgia.

SOLUTION: **So Answer D is not possible as the distance between Carlos and Georgia. Georgia cannot be more than 10 miles from Carlos (see Picture 2).**

SAMPLE TEST QUESTIONS

1 Michelle is trying to find the next number in this sequence:

1, 7, 13, 19, 25, 31, ___

Which of the following would most help Michelle to solve this problem?

A Make a model

B Draw a picture

C Break into parts

D Look for a pattern

2 Harold asked his friend George if he knows how many diagonals there are in a pentagon. Which of the following would help George the most in solving this problem?

A Act it out

B Guess and check

C Work backwards

D Draw a picture

3 Mrs. Jones asked 6 children to figure out how many handshakes there would be if all 6 children shook hands with each other. Which of the following would help most to solve this problem?

A Act it out

B Make a chart or graph

C Work backwards

D Guess and check

4 What is the area of the region enclosed by the figure below?

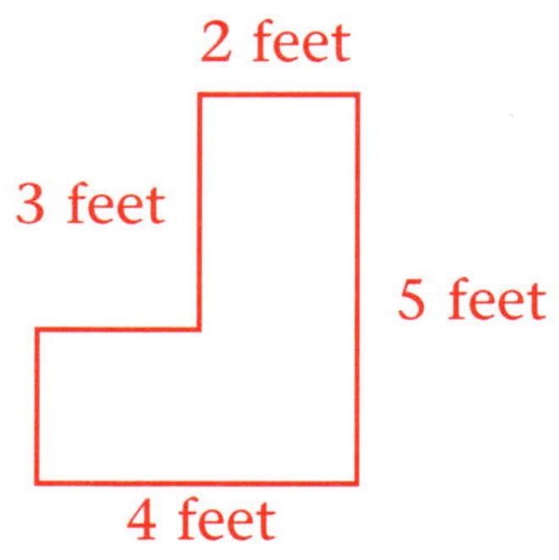

A 10 square feet

B 12 square feet

C 14 square feet

D 16 square feet

5 Samantha saves money every day. On Monday she saved $1, on Tuesday she saved $2, on Wednesday she saved $4, on Thursday she saved $8, and on Friday she saved $16. If she continues in the same way, how much money will she save on Saturday?

A $20

B $24

C $32

D $36

6 How many total squares are there in this grid?

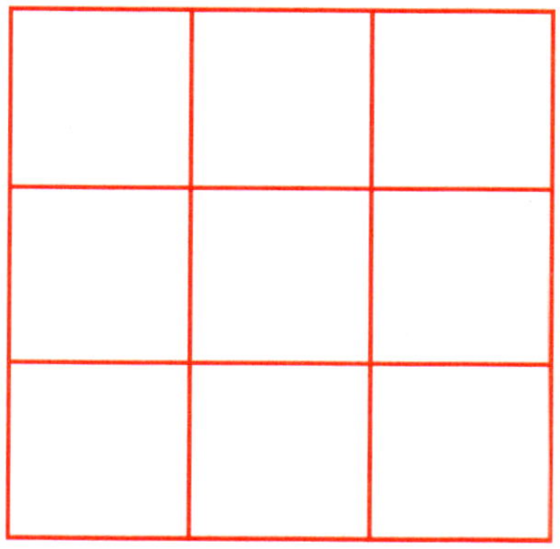

A 16

B 14

C 10

D 9

LESSON 35

SOLVING PROBLEMS IN MORE THAN ONE WAY

Some problems can be solved in more than one way. You may solve it one way and a friend may solve it another way.

EXAMPLE 1

Julie ran around the track 4 times. Each time around the track is 440 yards. To find how far she ran, Julie added: 440 + 440 + 440 + 440. What is another way to find how far she ran?

STRATEGY: **Read the problem again.**

SOLUTION: **Since she ran the same distance 4 times, another way to solve this problem is to multiply: 4 × 440.**

Writing a number sentence is often a good way to solve some problems.

EXAMPLE 2

A van from the mall cost \$2 for the first mile and \$0.75 for each mile after the first mile. How can you figure out how much it cost to travel 6 miles?

A 6 × \$0.75 **B** \$2 + 6 × \$0.75

C \$2 + 5 × \$0.75 **D** 5 × \$0.75

STRATEGY: **Think of the cost of the van in two parts, \$2 and \$0.75.**

The first mile cost \$2, and the other 5 miles cost \$0.75 each.

So the total cost is \$2 + the cost of the 5 miles.

SOLUTION: **The answer is \$2 + 5 × \$0.75, Answer C.**

SAMPLE TEST QUESTIONS

1 Melanie charges $4 an hour to babysit. Which of the following shows a way of finding how many hours she must work to earn $64?

A $64 \div 4$

B 64×4

C $64 - 4$

D $64 \div (2 + 4)$

2 It will take Dan 3 hours and 10 minutes to drive from home to the airport. It will take him another 40 minutes to get to the gate. He figures he needs $3\frac{1}{2}$ hours all together to get from his home to the gate. Is he right?

A No, because 3 hours and 10 minutes + 40 minutes = 4 hours.

B Yes, because 3 hours and 10 minutes + 40 minutes is less than $3\frac{1}{2}$ hours.

C Yes, because 3 hours and 10 minutes + 40 minutes = $3\frac{1}{2}$ hours.

D No, because 3 hours and 10 minutes + 40 minutes is more than $3\frac{1}{2}$ hours.

3 The cost of a special ice cream at the mall is $2.25. Kathy's father paid for 3 special ice creams. Kathy figured that the cost was $5.75. Was she correct?

A No, 2 × $2.25 = $4.50

B No, 3 × $2.25 = $6.75

C Yes, $8.00 – $2.25 = $5.75

D Yes, 3 + $2.75 = $5.75

4 Gwen waits 15 minutes every day for her ride. How long does she wait in 5 days all together?

To find out how long she waited, Gwen multiplied 5 × 15. What is another way to find how long she waited?

A $15 \div 5$

B $5 + 15$

C $15 + 15 + 15 + 15 + 15$

D $15 + 15 + 15$

Problems 5 through 8 can be solved in more than one way.

5 Kim is 5 feet 4 inches tall. Her brother is 4 feet 6 inches tall. How much taller is Kim than her brother?

A 6 inches

B 8 inches

C 10 inches

D 12 inches

6 Letters from A to F are worth 3 points; letters from G to M are worth 4 points; letters from N to S are worth 5 points; and letters from T to Z are worth 2 points. What is the value of "loyalty"?

A 25 points

B 22 points

C 21 points

D 20 points

7 Kelly gets 5 cents for each empty bottle she returns to the supermarket. How many bottles must she return to earn $5.50?

A 90

B 100

C 105

D 110

8 Karl spent $16 for a special model airplane. Jeff spent half as much for a model boat. How much did they spend all together?

A $24

B $20

C $16

D $8

REVIEW OF LESSONS 30-35

1 Add: $\frac{4}{15} + \frac{2}{15} = ?$

A $\frac{2}{15}$

B $\frac{2}{30}$

C $\frac{4}{15}$

D $\frac{6}{15}$

2 Subtract: 9.6 – 3.2 = ?

A 12.8

B 6.8

C 6.4

D 5.4

3 Find the missing number for this sentence.

16 – □ = 6

A 10

B 9

C 8

D 7

4 Samantha spent $25.46 at the mall. She started with $40. Which number sentence shows how much money she had left?

A 25.46 + 40 = ____

B 25.46 – 40 = ____

C 40 – 25.46 = ____

D 40 × 25.46 = ____

5 Dorothy earns $6.25 each hour she works at Columbia Computers Company. Don earns $7.50 each hour he works for the same company. If Don works 6 hours a day and Dorothy 7 hours a day, who earns more each day? How much more?

A Dorothy earns $1.75 more.

B Dorothy earns $1.25 more.

C Don earns $1.75 more.

D Don earns $1.25 more.

6 Smith has 18 dollars and Caroline has 20 dollars. How many nickels would that be all together?

Which number sentence could be used to solve this problem?

A $(18 + 20) \times 100$

B $(18 + 20) \times 20$

C $(18 + 20) \times 10$

D $(18 \times 20) + 20$

7 With each basketball game, Carlos scored more and more points. He scored 3 points in the first game, 7 points in the second game, and 11 points in the third game. If this pattern continues, how many points will Carlos score in the fifth game?

A 15

B 17

C 19

D 21

8 Patty says she can type 52 words per minute. At that rate, about how many words can she type in a half hour?

A 1,500

B 1,700

C 2,100

D 5,000

9 Jimmy said he took 5 steps forward and 3 steps backward every minute. After 20 minutes, how far did he get?

A 40 steps forward

B 20 steps forward

C 10 steps forward

D 10 steps backward

OPEN-ENDED QUESTIONS

1 Add 10 to each number.

a) 0.01 ______________

b) 0.1 ______________

c) 1 ______________

d) $1\frac{1}{2}$ ______________

e) 10.1 ______________

2 Add 0.1 to each number.

a) 10 ______________

b) 1 ______________

c) 0.1 ______________

d) 0.01 ______________

e) 10.1 ______________

3 What information is missing from this problem?

Max and Billie mowed lawns for 12 hours last week. They decided to split the money that they made evenly between the two of them. How much did each make?

Supply the missing information in this problem and then solve the problem. Show your work.

4 A spelling bee awards 3 points for spelling challenge words correctly and 1 point for regular words. Rhonda was given 11 words to spell and got 23 points. What combination of challenge and regular words did she spell correctly? Show your work.

LESSON 36

CLASSIFICATION

There are times when you have to sort objects that have the same shape, size, or color. When you do this you are classifying objects.

EXAMPLE 1

Which of the following best describes this group of robots?

A All the robots have 2 legs and only 1 eye.

B All the robots have 2 antennas and 2 legs.

C All the robots have 2 legs and 2 eyes.

D All the robots have 2 antennas and 2 eyes.

STRATEGY: **Write the number of antennas, eyes, and legs right on each robot.**

STEP 1: Count the number of antennas on each drawing.

Robot 1—2 antennas

Robot 2—1 antenna

Robot 3—2 antennas

Robot 4—1 antenna

Write these numbers on each robot, next to the antennas.

STEP 2: Count the number of eyes on each robot.

Robot 1—2 eyes

Robot 2—2 eyes

Robot 3—2 eyes

Robot 4—2 eyes

Write these numbers on each robot, next to the eyes.

STEP 3: Count the number of legs on each robot.

Robot 1—2 legs

Robot 2—2 legs

Robot 3—2 legs

Robot 4—2 legs

Write these numbers on each robot, next to the legs.

STEP 4: Check each answer against the numbers on the robots.

SOLUTION: **The answer is C. All robots have 2 legs and 2 eyes.**

SAMPLE TEST QUESTIONS

1 What pattern was used to sort these shapes into groups?

A large things and small things

B triangles and squares

C circles and figures with 4 sides

D triangles and figures with 4 sides

2 What would be a good rule for sorting these objects into exactly 2 groups?

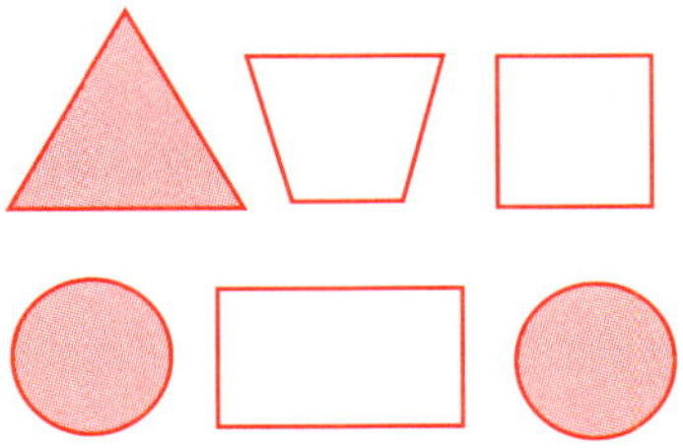

A circles in one group, rectangles in another group

B circles in one group, shapes with four sides in the other

C large shapes in one group, small shapes in the other

D shaded shapes in one group, unshaded shapes in the other

3 Which of the following would be the quickest way to count these squares?

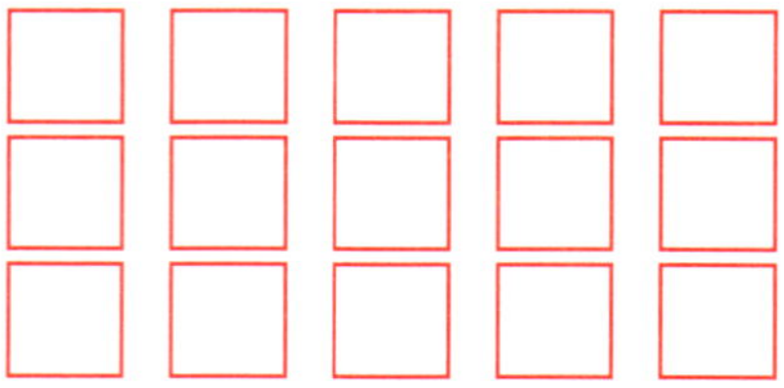

A by 1's

B by 2's

C by 3's

D by 4's

4 Which of these shapes would not be grouped with the others?

A

B

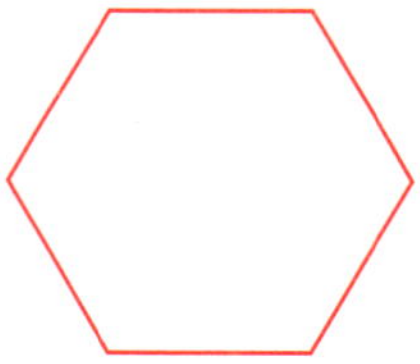

C

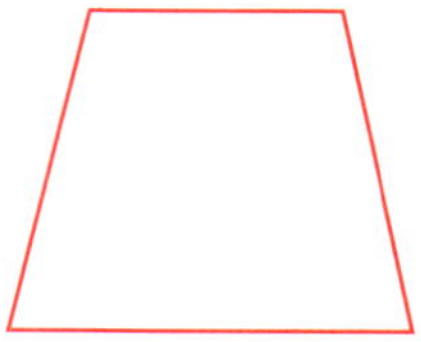

D

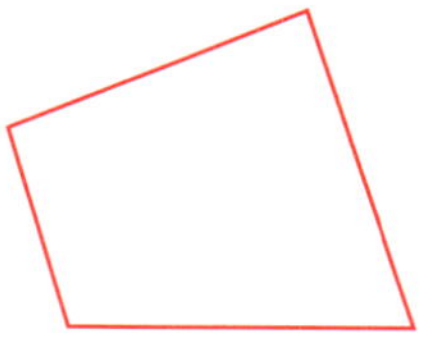

5 Which of the following best describes this group of Floxes?

A All Floxes have 2 legs and only 1 eye.

B All Floxes have 2 antennas and 2 legs.

C All Floxes have only 1 antenna and 2 eyes.

D All Floxes have 2 antennas and only 1 eye.

LESSON 37

ARRANGEMENTS AND PROBABILITY

ARRANGEMENTS

A good way to find the number of different **arrangements** (ways to put things in order) is to make a list of all possible arrangements.

EXAMPLE 1

Janell, Teresa, and Holly want to sit next to each other in the movies. How many different ways can they sit next to each other?

STRATEGY: **Make a list of all the arrangements.**

STEP 1: Use abbreviations for the names.

J for Janell

T for Teresa

H for Holly

STEP 2: List all possible seating arrangements.

J-T-H	J-H-T
T-J-H	T-H-J
H-J-T	H-T-J

STEP 3: Count all the arrangements.

SOLUTION: **There are 6 different ways the girls can sit next to each other.**

PROBABILITY

The probability of an event tells us how likely it is for the event to take place.

Probability is written as a number between 0 and 1, usually a fraction.

EXAMPLE 2

Suppose you have a bag of 10 marbles—2 are red, 3 are blue, and 5 are yellow. You put your hand in the bag and pick a marble. What color would it most likely be?

STRATEGY: **Follow these steps.**

STEP 1: Note the total number of marbles in the bag.

There are 10 marbles in the bag.

STEP 2: Figure out the chance of picking each color.

Since there are 2 red marbles, the chance of picking a red marble is 2 out of 10 or $\frac{2}{10}$.

(One way to think of this is to realize that if you were to stick your hand in 10 times and pull all the marbles out, 2 of them would be red. 2 red marbles out of 10 marbles gives us our probablility of 2 out of 10 or $\frac{2}{10}$.)

Since there are 3 blue marbles, the chance of picking a blue marble is 3 out of 10 or $\frac{3}{10}$.

Since there are 5 yellow marbles, the chance of picking a yellow marble is 5 out of 10 or $\frac{5}{10}$.

STEP 2: Choose the color whose chance of being picked is the highest.

This is yellow, with a chance of 5 out of 10 or $\frac{5}{10}$.

SOLUTION: **It would most likely be yellow.**

SAMPLE TEST QUESTIONS

1 How many different 3-digit numbers can be made using the digits 1, 2, and 3? (No digit can be repeated.)

A 4

B 6

C 8

D 10

2 There are 14 marbles in a bag. The marbles come in 4 colors. This table tells you the number of each color.

COLOR	NUMBER
Red	3
Blue	5
Green	2
Yellow	4

If you pick a marble from the bag without looking, what color are you most likely to get?

A red

B blue

C green

D yellow

3 If you spin the spinner, on what name is it most likely to stop?

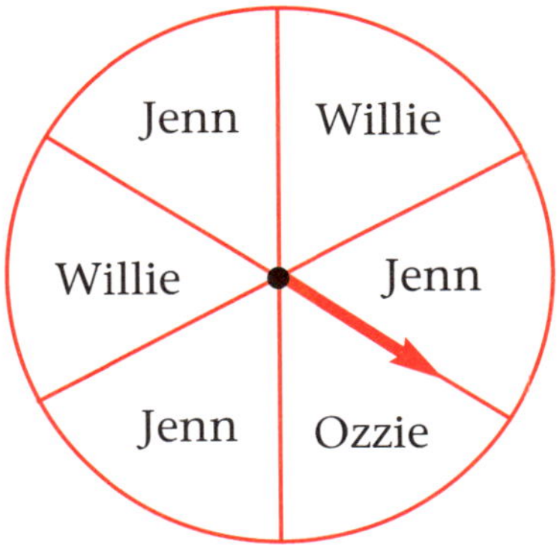

A Willie

B Jenn

C Ozzie

D Dominic

4 If you close your eyes and pick one of these figures, which figure are you least likely to pick?

A square

B rectangle

C circle

D triangle

5 How many different outfits can be made with 3 hats, 4 scarves, and 2 blouses?

A 9

B 18

C 24

D 32

6 How many different three-digit numbers can be made that use all three of these digits: 4, 4, and 5?

A 3

B 4

C 5

D 6

7 There are 3 juices, 5 sandwiches, and 4 desserts available at Arthur's school cafeteria. How many different meals can Arthur choose consisting of one juice, one sandwich, and one dessert?

A 12

B 19

C 30

D 60

8 Mandy placed 12 cards in a box. The cards had different names on them, except for two which had her brother's name Joseph on them. What is the probability that if Joseph reached into the bag without looking, he would pick his own name?

A $\frac{1}{12}$

B $\frac{1}{6}$

C $\frac{5}{6}$

D $\frac{11}{12}$

REVIEW OF LESSONS 36-37

1 Matthew has 4 new books he wants to place next to each other on his book shelf. How many different arrangements of these books are possible?

A 24 **B** 16

C 12 **D** 6

2 Which of the following best describes the group of Vikings?

A All the Vikings have only one eye and a spear.

B All the Vikings have two legs and a hat with two horns.

C All the Vikings have two legs and a hat with at least one horn.

D All the Vikings have a spear.

3 What is the probability that a coin will land on heads if you toss it once?

A $\frac{1}{3}$ **B** $\frac{1}{2}$

C $\frac{2}{3}$ **D** $\frac{1}{4}$

4 Which shape would not be grouped with the others?

A

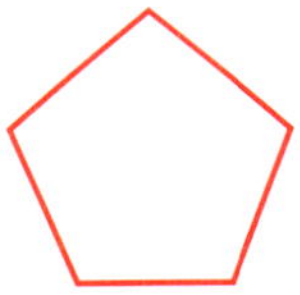

B

C

D

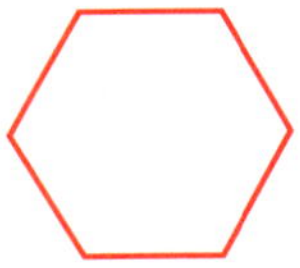

OPEN-ENDED QUESTIONS

1 Circle the shape that would not be grouped with the other three.

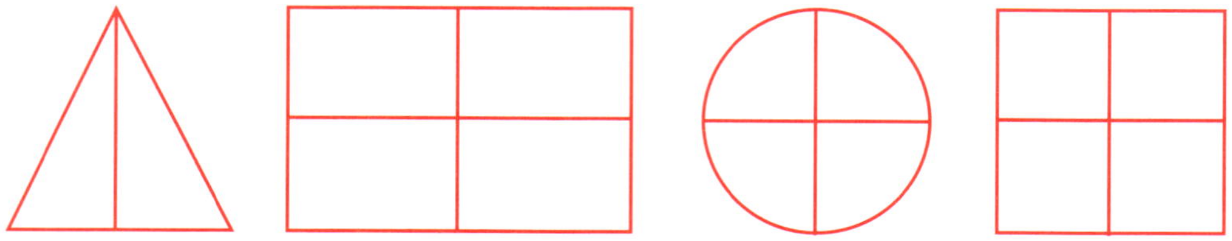

Write a sentence that tells why you circled that shape.

2 Kayla has a cube with the numbers 1 through 6 on the six faces.

a) What is the probability that the number 3 will come up if she rolls the cube? Explain your answer.

b) What is the probability that an even number will come up if she rolls the cube? Explain your answer.

MEAP MATHEMATICS, GRADE 4 PRACTICE TEST

1 Which of the following best describes the group of numbers: **5, 17, 35, 403, 899**?

A They are all even.

B They are all odd.

C They are a mixture of even and odd numbers.

D They all can be divided evenly by 5.

2 Which figure is $\frac{3}{5}$ shaded?

A

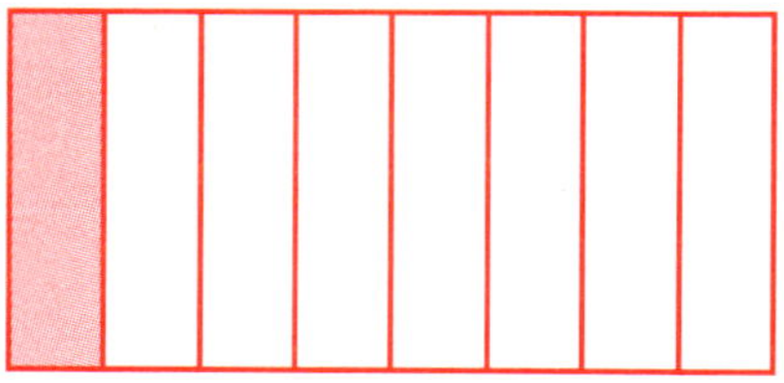

B

C

D

3 Which number is between 8 and 9?

A $3\frac{8}{9}$ **B** $6\frac{9}{10}$

C $7\frac{1}{10}$ **D** $8\frac{1}{2}$

4 What is the number pair that shows the location of G?

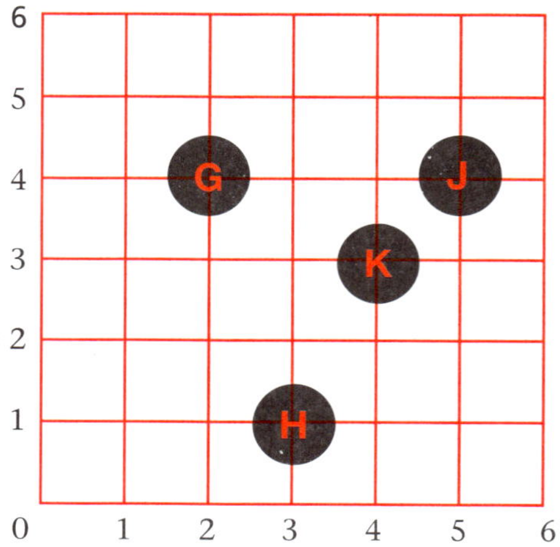

A (2,4)

B (2,1)

C (3,1)

D (4,3)

5 At which temperature will water boil?

A 0˚C

B 32˚C

C 100˚C

D 100˚F

6 Which statement is not true?

A $16 + 0 = 16$

B $0 + 0 = 0$

C $23 + 16 = 16 + 23$

D $23 + 0 = 0$

7 What is the decimal equivalent for $\frac{7}{100}$?

A 0.7 **C** 1.07

B 1.7 **D** 0.07

8 A very old car on Jeff's street sold for $7,604. Round this figure to the nearest thousand dollars.

A $8,000

B $7,500

C $7,000

D $6,000

9 What number is in the 10,000's place in 473,651?

A 4 **B** 7

C 3 **D** 6

10 When Jill passed the $1\frac{3}{4}$ mile mark of the race, her friend Pat was at the $1\frac{1}{4}$ mile mark. How far ahead of Pat was Jill?

A $\frac{1}{2}$ mile

B $\frac{1}{4}$ mile

C $1\frac{1}{4}$ miles

D $1\frac{1}{2}$ miles

11 Which figure has exactly two lines of symmetry?

A circle

B square

C triangle

D rectangle (not a square)

12 Which two figures are congruent?

A 2 and 3

B 4 and 6

C 2 and 4

D 1 and 7

13 How many faces does this figure have?

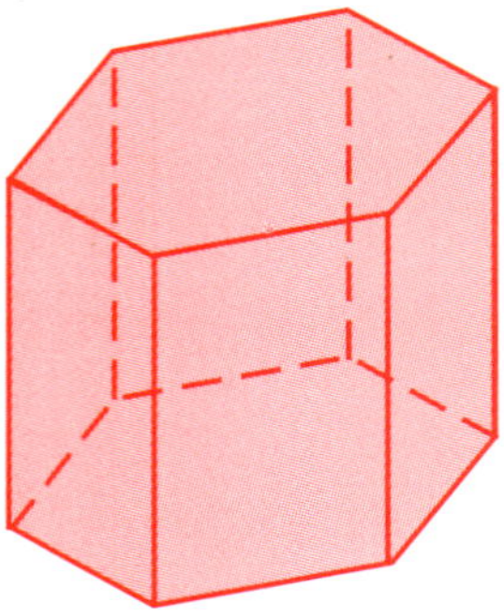

A 8 **B** 6

C 5 **D** 4

14 What would this figure look like if it were flipped to the right?

A 

B 

C 

D 

15 Which letter appears to identify the midpoint of line segment ST?

A S **B** T

C V **D** W

16 Which of the following shows a set of perpendicular lines?

A 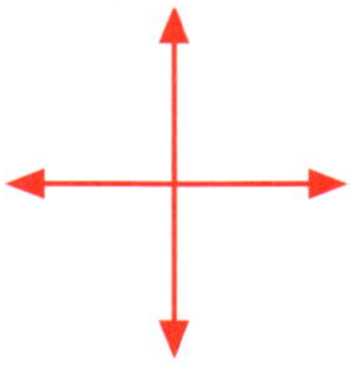

B 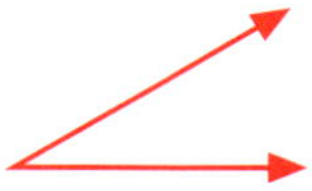

C 

D 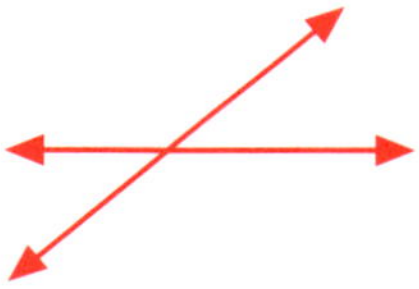

17 What kind of angle is this?

A obtuse angle

B acute angle

C right angle

D straight angle

18 Which of the following best describes railroad tracks?

A intersecting lines

B parallel lines

C perpendicular lines

D diagonal lines

19 Which is the next figure in the pattern?

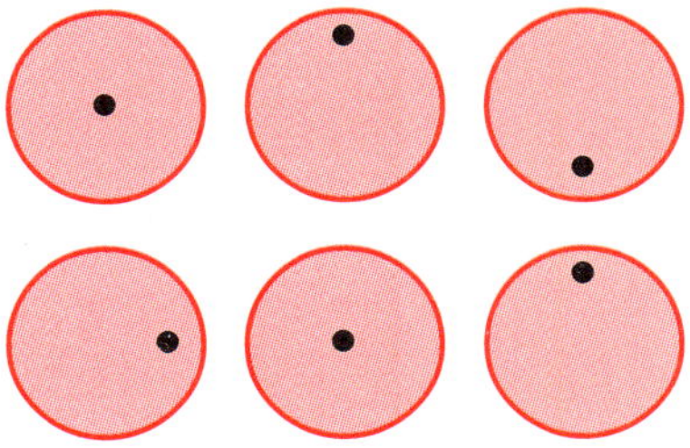

A

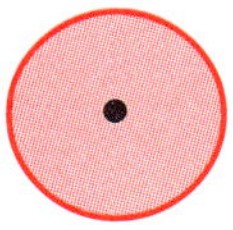

B

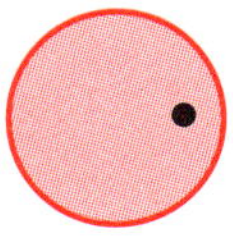

C

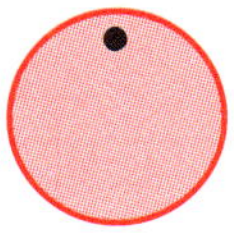

D

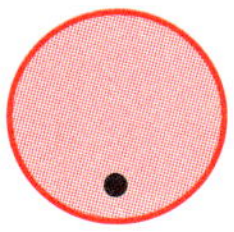

20 What number comes next in this pattern?

20, 21, 19, 20, 18, 19, 17, ____

A 17

B 18

C 19

D 20

21 Which two numbers should come next in the columns?

X	Y
4	1
8	2
12	3
—	—

A (16,5)

B (16,4)

C (12,4)

D (16,3)

22 Holly's dog weighs 6 kilograms. How many grams is this equal to?

A 6 **B** 60

C 600 **D** 6,000

23 What number is needed to make this sentence true?

5 × 2 + 3 = 7 + ___

A 6 **B** 8

C 10 **D** 12

24 What would be the best unit to measure the length of a football field?

A millimeter

B centimeter

C meter

D kilometer

25 Which of these is usually sold by the pound?

A milk

B orange juice

C apples

D cars

26 What is the area of the shaded region shown in this diagram?

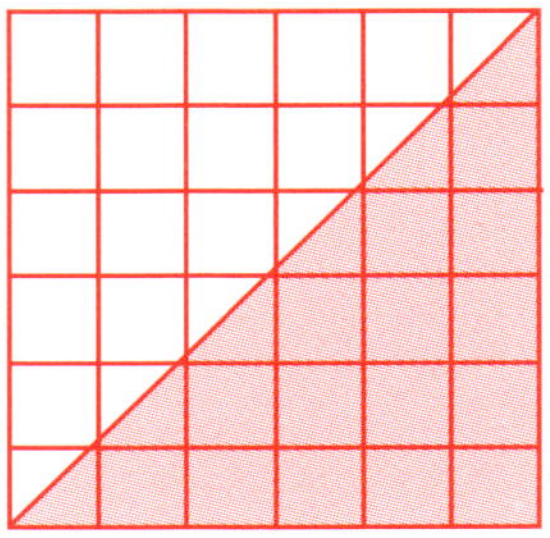

A 36 square units

B 24 square units

C 18 square units

D 15 square units

27 A quart of milk is the same as how many cups?

A 8 **B** 4

C 2 **D** 1

28 Angela spent $89.25 at the mall and $107.51 at the supermarket. Which is the closest estimate to how much she spent all together?

A $180 **B** $200

C $220 **D** $240

29 Melissa bought two videos at \$19 each and a CD-ROM disc for her computer at \$35. She then had \$7 left. How much money did she start with?

To find out how much money Melissa had at the beginning, you should do which of the following?

A draw a picture

B act it out

C work backward

D make a graph

30 Find $\frac{1}{3}$ of 87.

A 29 **B** 27

C 25 **D** 23

31 Admission to the amusement park costs \$12 during the week and \$18 on weekends. Sarah bought 4 tickets for weekdays and 3 tickets for weekends. How will she figure out how much money she needs?

A $(12 \times 4) + (18 \times 3)$

B $(12 + 4) + (18 + 3)$

C $(12 \times 3) + (18 \times 4)$

D $(12 + 18) \times (3 + 4)$

32 Which type of graph is shown below?

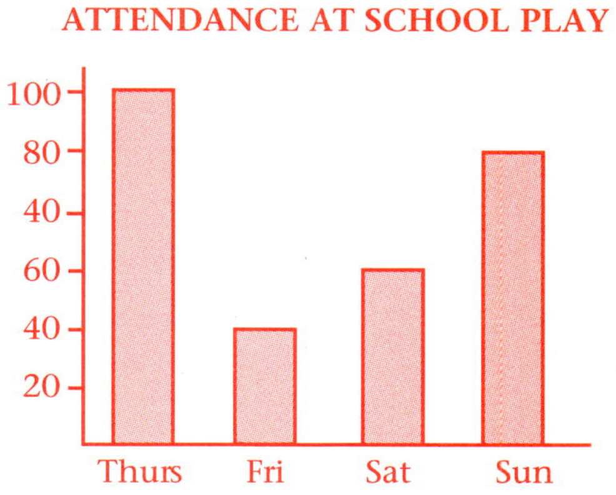

A line graph

B bar graph

C circle graph

D pictograph

33 The graph shows the favorite subjects of students in the fourth grade at the Star School.

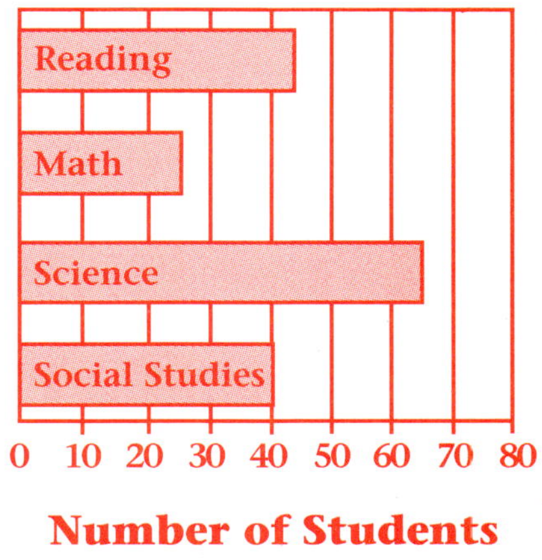

How many more students chose social studies more than math?

A About 40

B About 25

C About 15

D About 10

34 How many different 2-digit numbers can you make by using the digits 1, 3, and 5? (You cannot use a digit twice, such as 33.)

A 4

B 6

C 7

D 8

35 Jane is one of 14 people at a birthday party. The names of two people will be selected from a box to win special prizes. What are the chances that Jane will win a prize?

A $\frac{1}{14}$ **B** $\frac{1}{7}$

C $\frac{2}{7}$ **D** $\frac{6}{7}$

36 Jack saved $10 at the end of the first week of the month. At the end of each week, he doubled the amount of money he saved and added it to the amount from the previous week. After 4 weeks how much money did he save?

A $80 **B** $100

C $150 **D** $200

37 The taxi costs $1.25 for each mile. How much should a taxi ride of 4 miles cost?

A $5 **B** $6

C $8 **D** $10

38 Choose the number that must be removed to make this a true sentence.

16 + 17 + 18 + 19 = 52

A 16 **B** 17

C 18 **D** 19

39 Which sign makes this a true sentence?

6 × 7 – 14 ◯ 10 + 7

A = **B** <

C > **D** ×

40 Start at ▲; then go 2 blocks north; then go 1 block east; and 3 blocks south. Where do you end up?

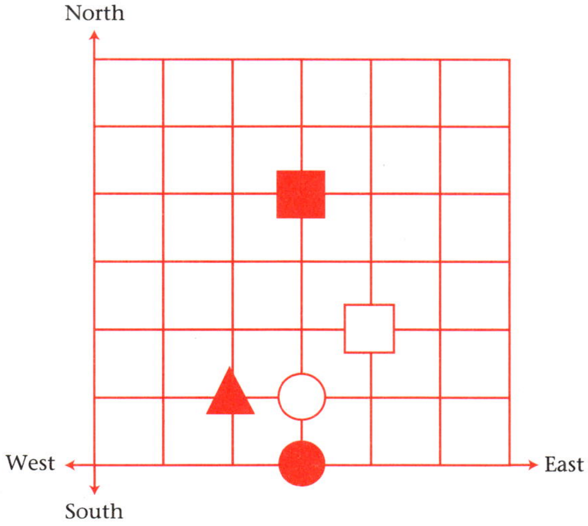

A ■ **B** □

C ● **D** ○

41 Joanne has a toy train that is 200 cm long. How many meters long is the train?

A 1 meter

B 2 meters

C 4 meters

D 20 meters

42 How much time has passed since 6:00?

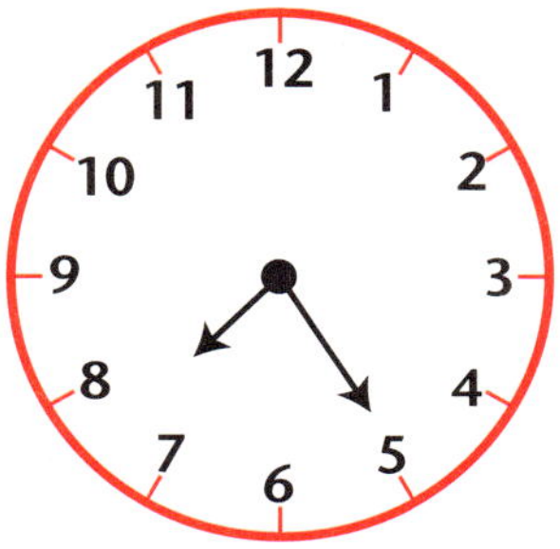

A 1 hour

B 1 hour, 15 minutes

C 1 hour, 25 minutes

D 1 hour, 35 minutes

43 Add: $\frac{4}{7} + \frac{2}{7} =$

A $\frac{6}{7}$

B $\frac{8}{7}$

C $\frac{2}{7}$

D $\frac{6}{14}$

44 Karl and Tony went to the football game. They left at 10:30 a.m. and returned at 6:00 p.m. How long were they away?

A 6 hours

B $7\frac{1}{2}$ hours

C $9\frac{1}{2}$ hours

D 10 hours

OPEN-ENDED QUESTIONS

45 The following information about how many students own scooters was collected in Brianna's school.

4th graders	2
5th graders	13
6th graders	27
7th graders	30
8th graders	22

Use tally marks in this chart to show the information above.

Grade	How Many Own a Scooter

46 Draw a picture of a circular cake and show how it can be cut to have 8 pieces of the same size. Then shade in $\frac{1}{4}$ of the pieces.

47 Draw a picture of each of these figures and complete the table about those figures. The information about the triangle is done for you.

Name of Figure	Drawing of Figure	Number of Sides	Number of Vertices
Triangle		3	3
Rectangle			
Hexagon			
Square			

48 Draw two different rectangles on the grid with an area of 18 squares.

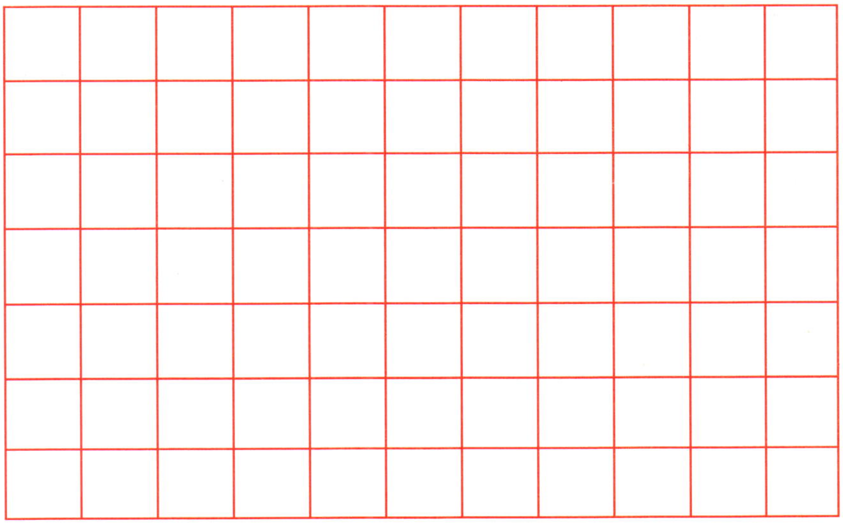

Find the perimeter of each of the two rectangles.

49 Solve this problem and show all your work.

Walt has only 5-dollar bills and 10-dollar bills in his wallet. If he pays $45, what combinations of 5- and 10-dollar bills could he use?